Tests and Quizzes

Precalculus
Graphical, Numerical, Algebraic
Seventh Edition

Franklin D. Demana
The Ohio State University

Bert K. Waits
The Ohio State University

Gregory D. Foley
The Liberal Arts and Science Academy of Austin

Daniel Kennedy
Baylor School

Boston San Francisco New York
London Toronto Sydney Tokyo Singapore Madrid
Mexico City Munich Paris Cape Town Hong Kong Montreal

⚠ This work is protected by United States copyright laws and is provided solely for the use of instructors in teaching their courses and assessing student learning. Dissemination or sale of any part of this work (including on the World Wide Web) will destroy the integrity of the work and is not permitted. The work and materials from it should never be made available to students except by instructors using the accompanying text in their classes. All recipients of this work are expected to abide by these restrictions and to honor the intended pedagogical purposes and the needs of other instructors who rely on these materials.

Reproduced by Pearson Addison-Wesley from electronic files supplied by the author.

Copyright © 2007 Pearson Education, Inc.
Publishing as Pearson Addison-Wesley, 75 Arlington Street, Boston, MA 02116.

All rights reserved. This manual may be reproduced for classroom use only. Printed in the United States of America.

ISBN   0-321-36992-0

1 2 3 4 5 6 BB 09 08 07 06

Contents

CHAPTER P   **Prerequisites**
- Quiz P.1–P.4   1
- Quiz P.5–P.7   2
- Chapter Test Form A   3
- Chapter Test Form B   5

CHAPTER 1   **Functions and Graphs**
- Quiz 1.1–1.3   7
- Quiz 1.4–1.7   8
- Chapter Test Form A   9
- Chapter Test Form B   11

CHAPTER 2   **Polynomial, Power, and Rational Functions**
- Quiz 2.1–2.4   13
- Quiz 2.5–2.8   15
- Chapter Test Form A   17
- Chapter Test Form B   19

CHAPTER 3   **Exponential, Logistic, and Logarithmic Functions**
- Quiz 3.1–3.4   21
- Quiz 3.5–3.6   23
- Chapter Test Form A   25
- Chapter Test Form B   27

CHAPTER 4   **Trigonometric Functions**
- Quiz 4.1–4.4   29
- Quiz 4.5–4.8   30
- Chapter Test Form A   32
- Chapter Test Form B   34

CHAPTER 5   **Analytic Trigonometry**
- Quiz 5.1–5.3   36
- Quiz 5.4–5.6   38
- Chapter Test Form A   40
- Chapter Test Form B   42

Midterm Exam P–5
- Midterm Exam A   44
- Midterm Exam B   48

CHAPTER 6 — Applications of Trigonometry

Quiz 6.1–6.3	52
Quiz 6.4–6.6	54
Chapter Test Form A	56
Chapter Test Form B	59

CHAPTER 7 — Systems and Matrices

Quiz 7.1–7.2	62
Quiz 7.3–7.5	64
Chapter Test Form A	66
Chapter Test Form B	69

CHAPTER 8 — Analytic Geometry in Two and Three Dimensions

Quiz 8.1–8.3	72
Quiz 8.4–8.6	73
Chapter Test Form A	75
Chapter Test Form B	77

CHAPTER 9 — Discrete Mathematics

Quiz 9.1–9.3	79
Quiz 9.4–9.6	81
Quiz 9.7–9.8	83
Chapter Test Form A	85
Chapter Test Form B	87

CHAPTER 10 — An Introduction to Calculus: Limits, Derivatives, and Integrals

Quiz 10.1–10.4	89
Chapter Test Form A	90
Chapter Test Form B	93

Final Exam 6–10

Final Exam A	96
Final Exam B	101

Answers   107

P.1–P.4

P Chapter Quiz

1. Write the inequality that describes the graph.

2. Solve $3(x + 2) = 5(2x - 3) - 7$.

3. Find the distance between the points $(-8, 5)$ and $(3, 2)$.

4. Simplify the expression $\dfrac{u^2 v^{-3}}{(u^3 v)^{-1}}$. Assume that the variables are not zero.

   A. $\dfrac{u^5}{v^2}$ **B.** $\dfrac{1}{uv^4}$ **C.** $\dfrac{1}{uv^2}$ **D.** $u^5 v^4$ **E.** $u^5 v^2$

5. In the xy-plane, graph $y = \dfrac{1}{2}x - 2$.

   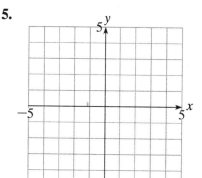

6. Solve $3x - 4 > -2$ algebraically.

7. In the xy-plane, what is the slope and y-intercept of the graph of the line $4x + 3y = 12$?

   A. $m = 4, b = 12$ **B.** $m = \dfrac{3}{4}, b = 3$

   C. $m = \dfrac{3}{4}, b = 4$ **D.** $m = -\dfrac{4}{3}, b = 3$

   E. $m = -\dfrac{4}{3}, b = 4$

8. Find the equation of the line that passes through the point $(10, -3)$ and whose graph is perpendicular to the graph of $y = 5x - 2$.

9. Find the midpoints of the sides of the triangle defined by the vertices $(5, 5)$, $(-2, 7)$, and $(-1, 1)$.

10. Solve $-6 \leq \dfrac{2x + 1}{4} \leq 6$ algebraically.

P Chapter Quiz

1. Which of the following are solutions of the equation $(x-5)(x+3) = 0$?
   A. $x = 5$ only
   B. $x = -3$ only
   C. Both $x = 5$ and $x = -3$
   D. Neither $x = 5$ nor $x = -3$
   E. There are no solutions to the equation.

2. Solve $2x^2 + 7x - 15 = 0$ using the quadratic formula.

3. Solve the inequality $|2x + 1| < 5$ algebraically.

4. Which of the following is the graph of $x = -5$ in the xy-plane?
   A. A line parallel to the x-axis five units above the x-axis
   B. A line parallel to the x-axis five units below the x-axis
   C. A line parallel to the y-axis five units to the right of the y-axis
   D. A line parallel to the y-axis five units to the left of the y-axis
   E. A line through the origin and the point $(5, 5)$

5. Solve: $\left|\dfrac{x+1}{2}\right| > 4$.

6. Solve the equation $x^2 + 14x = -9$ by completing the square.

7. Write the complex number $\dfrac{(3-2i)^2(3+2i)}{(5-i)}$ in standard form.

8. Solve: $x^2 - 2x - 2 < 1$.

9. Solve: $x^3 + 2x^2 + x \geq 0$.

10. Draw the graph of $y = 2x^2 - x - 3$. On the graph show the solutions to the quadratic equation $2x^2 - x - 3 = 0$.

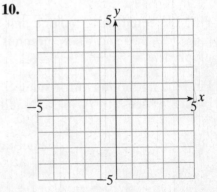

FORM A DATE _____

P Chapter Test   NAME _____

Directions: Show all work where appropriate. A graphing calculator may be necessary to answer some questions.

1. Name the algebraic property or properties illustrated by the equation.
   $3(x + y) = 3x + 3y$

   1. _____

2. Complete the table of solutions for the equation $y = 2x - 5$.

   2.
x	y
-2	
3	
	-7
	3

3. This year a news agency purchased a delivery vehicle for $27,000. The vehicle's value will depreciate $4,500 a year for each of the next 6 years. The equation $y = 27{,}000 - 4{,}500x$, $0 \le x \le 6$ models the vehicle's value, where y is the value of the vehicle and x is the number of years after purchase. In how many years will the value of the vehicle be $13,500?

   3. _____

4. Write the complex number $\dfrac{7 + 4i}{4 - 3i}$ in standard form.

   4. _____

5. For the function $f(x) = x^2 + 5x - 6$, choose the viewing window that shows two intersections with the x-axis and all the points of the graph in between.
   A. $[-10, 10]$ by $[-10, 10]$
   B. $[-2, 7]$ by $[-20, 20]$
   C. $[-5, 5]$ by $[-15, 15]$
   D. $[-7, 2]$ by $[-20, 20]$
   E. $[-20, 20]$ by $[-10, 10]$

   5. _____

6. Find the equation of a line that contains $(3, 4)$ and has slope 2. Write the equation in point-slope form.

   6. _____

7. Find the equation of a line that has x-intercept 2 and y-intercept -7. Write the equation in slope-intercept form.

   7. _____

8. In standard form, write the equation of a circle with center $(2, -6)$ and radius 9.

   8. _____

FORM A

P Chapter Test (continued)   NAME

9. Solve graphically:
$|6 - 4x| = 5$

9. _____

10. Solve graphically:
$4x^3 - 2x^2 - 4x - 1 = 0$

10. _____

11. Write the equation of a line through the point $(-1, 2)$ and parallel to the line with equation $3x - 2y - 5 = 0$. Write the equation in the general form.
A. $3x - 2y + 2 = 0$
B. $2x + 3y - 4 = 0$
C. $2x + 3y - 5 = 0$
D. $3x - 2y + 7 = 0$
E. $2x - 3y - 4 = 0$

11. _____

12. Solve the inequality $|2x + 3| > 1$ algebraically. Write the solution in interval notation and draw its number line graph.

12. _____

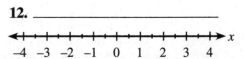

13. Use the quadratic formula to solve $6x^2 - 4x + 5 = 0$.

13. _____

14. Solve the inequality $x^2 - x - 6 \leq 0$. Express your answer in interval notation.

14. _____

15. Solve the equation $x^2 + 5 = 8x$ by completing the square. Show all steps.

15. _____

16. A projectile is launched straight up from the ground with an initial velocity of $320 \frac{\text{ft}}{\text{sec}}$. At what time(s) will the projectile be at least 1,584 feet above the ground?

16. _____

FORM B
DATE
P Chapter Test
NAME

Directions: Show all work where appropriate. A graphing calculator may be necessary to answer some questions.

1. Name the algebraic property or properties illustrated by the equation.
$2x + 3y = 3y + 2x$

1. _____

2. Complete the table of solutions for the equation $y = -3x + 2$.

2.

x	y
-2	
3	
	-4
	5

3. Last week a theater purchased an air conditioner for $24,000. The air conditioner's value will depreciate $4,000 a year for each of the next 6 years. The equation $y = 24,000 - 4,000x, 0 \le x \le 6$ models the air conditioner's value, where y is the value of the air conditioner and x is the number of years after purchase. In how many years will the value of the air conditioner be $8,000?

3. _____

4. Write the complex number $\dfrac{6 - 5i}{4 + 2i}$ in standard form.

4. _____

5. For the function $f(x) = x^2 + 5x + 6$, choose the viewing window that shows two intersections with the x-axis and all the points of the graph in between.
   A. $[-10, 10]$ by $[-10, 10]$
   B. $[-2, 7]$ by $[-20, 20]$
   C. $[-5, 5]$ by $[-5, 5]$
   D. $[-7, 2]$ by $[0, 20]$
   E. $[-20, 20]$ by $[0, 10]$

5. _____

6. Find the equation of a line that contains $(1, 3)$ and has slope -4. Write the equation in point-slope form.

6. _____

7. Find the equation of a line that has x-intercept 3 and y-intercept -5. Write the equation in slope-intercept form.

7. _____

8. In standard form, write the equation of a circle with center $(-3, 4)$ and radius 7.

8. _____

FORM B

P Chapter Test (continued)   NAME

9. Solve graphically:
$|3 + 5x| = 7$

9. _____

10. Solve graphically:
$-3x^3 - 2x^2 + 4x - 1 = 0$

10. _____

11. Write the equation of a line through the point $(-1, 2)$ and perpendicular to the line with equation $2x - 3y - 5 = 0$. Write the equation in the general form.
   A. $2x + 3y + 8 = 0$
   B. $3x + 2y + 2 = 0$
   C. $2x - 3y + 8 = 0$
   D. $3x + 2y - 1 = 0$
   E. $2x - 3y - 5 = 0$

11. _____

12. Solve the inequality $|3x - 4| \geq 2$ algebraically. Write the solution in interval notation and draw its number line graph.

12. _____

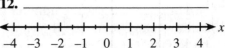

13. Use the quadratic formula to solve $4x^2 + 6x + 7 = 0$.

13. _____

14. Solve the inequality $x^2 + 3x - 4 \leq 0$. Express your answer in interval notation.

14. _____

15. Solve the equation $x^2 + 4 = 9x$ by completing the square. Show all steps.

15. _____

16. A projectile is launched straight up from the ground with an initial velocity of $352 \frac{\text{ft}}{\text{sec}}$. At what time(s) will the projectile be at least 1,536 feet above the ground?

16. _____

1.1–1.3 DATE

1 Chapter Quiz   NAME

1. Find the domain of the function $f(x) = x^2 + \sqrt{x-5}$.   1. _____
   Give your answer in interval notation.

2. Find the range of the function $f(x) = x^3 + \sqrt{x-2}$.   2. _____
   Give your answer in interval notation.

3. Suppose the point $(3, -5)$ lies on a graph of an even   3. _____
   function. Determine a second point on the graph.

4. Give an example of a discontinuous function and state   4. _____
   why it is discontinuous.

5. A rectangular field is to be enclosed by a fence. An   5. _____
   existing fence will form one side of the enclosure. The
   amount of new fence bought for the other three sides is
   1200 feet. What is the maximum area of the enclosed
   field?

6. Solve $100x^2 + 13x = 828$ by using the quadratic   6. _____
   formula. Give your answer to one decimal place.

7. Solve $\sqrt{x+2} = x - 4$ algebraically and support   7. Solution(s): _____
   graphically. Identify all extraneous solutions.

   Extraneous: _____

8. (a) Find all the zeroes for the function   8. (a) _____
   $f(x) = 3x^4 - 6x^2$.

   (b) List the intervals where the function   (b) _____
   $f(x) = 3x^4 - 6x^2$ is increasing.

9. The graph of which of the following functions is   9. _____
   bounded above by $x = 2$?

   A. $y = \dfrac{2x}{x^3 + 2}$ **B.** $y = \dfrac{2x}{x^2 + 1}$

   C. $y = \dfrac{2x^2}{x^2 + 1}$ **D.** $y = \dfrac{2x^3}{x^2 + 1}$

   E. $y = \dfrac{2x^4}{x^2 + 1}$

10. The graph to the right is a slight variation of one of the
    ten basic functions. Which of the following equations
    best represents the graph?

    A. $y = -\ln(x + 1)$ **B.** $y = -\ln(x - 1)$
    C. $y = -\sqrt{x} + 1$ **D.** $y = \sqrt{-x + 1}$
    E. $y = -\sqrt{x + 1}$

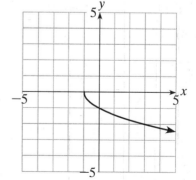

10. _____

1.4–1.7

DATE _____

1 Chapter Quiz

NAME _____

1. Which of these functions is one-to-one?

   A. $f(x) = 2x^3 - 3x$ B. $f(x) = 6x + 14$
   C. $f(x) = \sqrt{x^2 - 3x - 2}$
   D. $f(x) = 5x^2 - x + 3$
   E. $f(x) = \dfrac{x^2 + 7}{x - 1}$

1. _____

2. Graph the relation defined by the parametric equations $x = 2t^2 - 4$ and $y = 2t$. Use a $[-5, 5]$ by $[-5, 5]$ viewing window and $-2 \leq t \leq 2$.

2.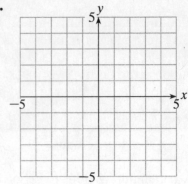

3. Describe how the graph of $f(x) = x^4 - 2$ can be transformed to the graph of $g(x) = (2x)^4 - 2$.

3. _____

4. Let $f(x) = 2x^3 - 3x^2 + 5x + 6$. Find an equation for g, the reflection of f across the y-axis.

4. _____

5. Write the equation whose graph can be obtained from the graph of $y = x^2$ by a vertical stretching of factor 4, a reflection through the x-axis, and a vertical shift 5 units up.

5. _____

6. Find the inverse relation of $f(x) = \dfrac{6x + 5}{3x - 1}$.

6. _____

7. What two functions implicitly define the relation $x + |y| = 2$? What is the domain?

7. _____

   Domain: _____

8. Alfred receives a salary of $2500 per month, and he pays $550 per month in rent. What percentage of his salary is spent on rent?

8. _____

9. Write an equation that models the following and solve. How many gallons of vanilla ice cream with 12% butterfat must be mixed with 20 gallons of chocolate ice cream with 20% butterfat to make vanilla fudge ice cream with 15% butterfat?

9. Equation: _____

   Solution: _____

10. Let A represent the amount of money Paul has in his pocket. Paul and his girl friend go out to dinner. They spend x dollars on food, pay 8% tax on the dinner check, and leave a 15% tip (not including the tax).

    (a) Write a function relating the amount of money in Paul's pocket to the amount they spend on food for dinner.

    (b) If Paul has $46 in his pocket, what is the maximum amount they could spend on food for dinner?

10. (a) _____

    (b) _____

FORM A

1 Chapter Test

DATE

NAMES

Directions: Show all work where appropriate. A graphing calculator may be necessary to answer some questions.

1. Determine the *domain* of the real-valued function $f(x) = \ln(4 - x)$.

   A. $(0, \infty)$ B. $(4, \infty)$ C. $(-\infty, 4)$
   D. $[4, \infty)$ E. $(-\infty, -4]$

   1. _____

2. Find all vertical and horizontal asymptotes of the graph of $y = \dfrac{3x}{x^2 - 3x - 10}$.

   2. Vertical: _____

   Horizontal: _____

3. (a) Graph the piecewise function
   $$f(x) = \begin{cases} x^2 + 4, & \text{if } x \leq 0 \\ 3x - 5, & \text{if } x > 0 \end{cases}.$$

   (b) Is the function continuous? Why or why not?

   3. (a)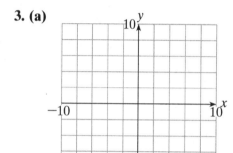

   (b) _____

4. Solve the equation $-x^2 - 3x + 10 = 0$. Give exact answers.

   4. _____

5. Juan drives in city traffic for 2 hours. He travels 62 miles during that time. He averages 26 miles per hour less on this drive than he does on a country highway. What is his average speed on the country highway?

   5. _____

6. Solve the inequality $|2(x - 2)| > 1$ algebraically. Write the solution in interval notation and draw its number line graph.

   6. _____

   ←+—+—+—+—+—+—+—+—+—→ x
   −4 −3 −2 −1 0 1 2 3 4

7. Let $f(x) = 3x^2 - 5$ and $g(x) = \dfrac{2}{x}$. Find the function $f \circ g$. Give the domain of $f \circ g$.

   7. _____

   Domain: _____

8. A cylindrical tank with diameter 30 ft is filled with gasoline to a depth of 60 ft. The gasoline begins draining at a constant rate of 5 cubic feet per second. Write the volume of gasoline remaining in the tank t seconds after the tank begins draining as a function of t.

   8. _____

FORM A

1 Chapter Test (continued)   NAME

9. Graph the relation defined by the parametric equations below. Use an appropriate window size for $-3 \leq t \leq 3$.

   $x = 3 - t^2,\ y = 1 + 2t$

9.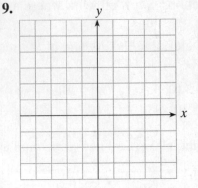

10. Describe how the graph of $y = (2x + 2)^2$ can be obtained from the graph of $y = x^2$.

10. _____

_____

_____

11. Three of the twelve basic functions are even functions. Which are they?

11. _____

_____

_____

12. A ball is hit straight up from a height of 3 ft with an initial velocity of 60 ft/sec. Write a set of parametric equations that you would use to graph height against time. Then find the maximum height of the ball and the number of seconds to reach that height.

12. $x =$ _____

$y =$ _____

Max. ht. = _____

Time = _____

13. Use a graphing calculator to determine all local maxima and/or minima for the function $y = 2x^3 - 7x^2 - 4x$. Give the values where the extremum occur to two decimal places.

13. _____

14. Let $f(x) = \sqrt{2 - x}$.

   (a) Why does f have an inverse that is a function?

   (b) Find a rule for $f^{-1}(x)$ and state its domain.

14. (a) _____

(b) _____

Domain: _____

15. What are the upper and lower bounds for the function $f(x) = \sin x + 2$?

15. Upper: _____

Lower: _____

FORM B

1 Chapter Test

Directions: Show all work where appropriate. A graphing calculator may be necessary to answer some questions.

1. Determine the *domain* of the real-valued function $f(x) = \ln(x - 4)$.

   A. $(0, \infty)$ B. $(4, \infty)$ C. $(-\infty, 4)$
   D. $(4, -\infty)$ E. $(-\infty, -4)$

   1. _____

2. Find all vertical and horizontal asymptotes of the graph of $y = \dfrac{5x}{2x^2 - 11x - 6}$.

   2. Vertical: _____

   Horizontal: _____

3. (a) Graph the piecewise function
$$f(x) = \begin{cases} x^3 + 2, & \text{if } x > 0 \\ 2 - x, & \text{if } x < 0 \end{cases}.$$

   (b) Is the function continuous? Why or why not?

   3. (a)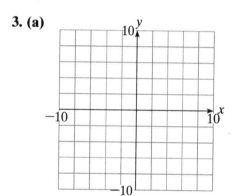

   (b) _____

4. Solve the equation $-x^2 - x + 12 = 0$. Give exact solutions.

   4. _____

5. Angela drives on a county highway for 2 hours. She travels 112 miles during this time. She averages 22 mph faster on this drive than she does in the city traffic. What is her average speed in the city?

   5. _____

6. Solve the inequality $|3x + 6| > 1$ algebraically. Write the solution in interval notation and draw its number line graph.

   6. _____

   ←——+——+——+——+——+——+——+——+——→ x
   −4 −3 −2 −1 0 1 2 3 4

7. Let $f(x) = -5(x + 1)^2$ and $g(x) = \dfrac{x - 2}{2}$. Find the function $f \circ g$. Give the domain of $f \circ g$.

   7. _____

   Domain: _____

8. A cylindrical tank with diameter 25 meters is filled with gasoline to a depth of 40 meters. The gasoline begins draining at a constant rate of 4 cubic meters per second. Write the volume of gasoline remaining in the tank t seconds after the tank begins draining as a function of t.

   8. _____

FORM B

1 Chapter Test (continued)   NAME

9. Graph the relation defined by the parametric equations below. Use an appropriate window size for $-3 \leq t \leq 3$.

   $x = 5 - 2t^2, y = 2(t - 1)$

9.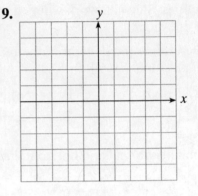

10. Describe how the graph of $y = (5x - 5)^2$ can be obtained from the graph of $y = x^2$.

10. _____
    _____
    _____

11. Five of the twelve basic functions are odd functions. Which are they?

11. _____
    _____
    _____

12. A ball is hit straight up from a height of 4 ft with an initial velocity of 65 ft/sec. Write a set of parametric equations that you would use to graph height against time. Then find the maximum height of the ball and the number of seconds to reach that height.

12. $x =$ _____
    $y =$ _____
    Max. ht. = _____
    Time = _____

13. Use a graphing calculator to determine all local maxima and/or minima for the function $y = 2x^3 - 5x^2 - 4x$. Give the values where the extremum occur to two decimal places.

13. _____

14. Let $f(x) = \sqrt{x - 5}$.

    (a) Why does f have an inverse that is a function?
    (b) Find a rule for $f^{-1}(x)$ and state its domain.

14. (a) _____
    (b) _____
    Domain: _____

15. What are the upper and lower bounds for the function $f(x) = \cos x - 1$?

15. Upper: _____
    Lower: _____

2.1–2.4

2 Chapter Quiz

DATE _____

NAME _____

1. Write an equation in standard form for the parabola that has vertex $(3, -2)$ and passes through the point $(1, 14)$.

   1. _____

2. Draw the graph of $f(x) = 0.05x^3 + 6x^2 - 2x - 3$ in the $[-15, 10]$ by $[-100, 175]$ viewing rectangle. How many real zeros are evident from this graph?

   A. 1   B. 2
   C. 3   D. 0
   E. Infinitely many

   2. _____

3. Describe the end behavior of the polynomial function $f(x) = -6x^3 + 2x^2 + 3x - 8$.

   3. $f(x) \to$ _____ as $x \to -\infty$;

   $f(x) \to$ _____ as $x \to \infty$

4. Use the Remainder Theorem to find the remainder when $x^3 - 6x^2 + 5x - 2$ is divided by $x - 6$.

   4. _____

5. Find a polynomial of degree 3 whose zeros are $-3, \dfrac{3}{2}, 2$.

   A. $2x^3 - x^2 - 15x - 18$
   B. $2x^2 + 3x - 9$
   C. $2x^2 - 7x + 6$
   D. $2x^3 - x^2 - 15x + 18$
   E. $2x^3 - 7x^2 - 15x + 18$

   5. _____

6. Use long division to find the remainder when $x^4 - 3x^2 + 5x - 1$ is divided by $x^2 - 3$.

   6. _____

7. Use synthetic division to divide $\dfrac{3x^3 - 2x^2 + 5x - 3}{x + 2}$.
   Summarize your results by writing a fraction equation.
   $\dfrac{3x^3 - 2x^2 + 5x - 3}{x + 2} =$ _____

   7. _____

8. A contractor purchases a new bulldozer for $45,000. After 15 years the bulldozer will be outdated and have no value. Write a linear equation giving the value V of the equipment during the 15 years it will be used, where t is the number of years after purchase.

   8. _____

2 Chapter Quiz (continued)

2.1–2.4

NAME

9. The formula $h = -16t^2 + v_0 t + s_0$ gives the height of an object tossed upward where v_0 represents the initial velocity, s_0 represents the initial height, and t represents time. A golf ball is hit straight up from the ground level with an initial velocity of 72 ft/sec. Find the maximum height that the ball reaches and the number of seconds it takes to reach that height.

9. Max. ht. = _____

Time = _____

10. The manager of 100 apartments knows that at $600 rent per month, all apartments will be rented. For each $25 increase, one apartment will not be occupied. Let x represent the number of $25 increases to the rent.

(a) Write the revenue as a function of x.
(b) What rent per unit will yield maximum revenue?
(c) What is the maximum revenue?

10. (a) _____

(b) _____

(c) _____

2.5-2.8

2 Chapter Quiz

DATE

NAME

1. Find the domain of $f(x) = \dfrac{x^3 + 5x^2 - 2}{x^2 - 2}$.

1. _____

A. $(-\infty, -2) \cup (-2, 2)$
B. $(-\infty, -2) \cup (-2, 2) \cup (2, \infty)$
C. $(-\infty, -\sqrt{2}) \cup (-\sqrt{2}, \sqrt{2})$
D. $(-\infty, -\sqrt{2}) \cup (-\sqrt{2}, \sqrt{2}) \cup (\sqrt{2}, \infty)$
E. $(-\infty, \infty)$

2. Find all rational zeros of $f(x) = 2x^3 - x^2 - 23x - 20$.

2. _____

3. Find all the zeros of $f(x) = x^4 - x^3 - x^2 - x - 2$.

3. _____

4. Write a linear factorization of $f(x) = x^3 + 6x - 7$.

4. _____

5. Which of the following gives the zeros of the graph and their multiplicity?

5. _____

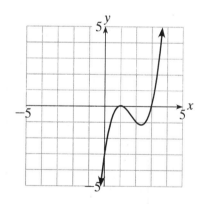

A. 1 (multiplicity 1), 3 (multiplicity 2)
B. 1 (multiplicity 3), 2 (multiplicity 1)
C. 1 (multiplicity 3), 3 (multiplicity 1)
D. 1 (multiplicity 2), 3 (multiplicity 1)
E. 1 (multiplicity 1), 2 (multiplicity 3)

6. Solve the inequality $\dfrac{x + 5}{|x - 2|} \leq 0$.

6. _____

2 Chapter Quiz (continued)   NAME

7. Solve the rational equation $\dfrac{x(2x+1)}{x-2} = \dfrac{10}{x-2} - \dfrac{5}{2}$.

7. Root: _____

 Extraneous root: _____

8. Find a polynomial of degree 2 with real-number coefficients and zero $3 - 2i$.

8. _____

9. Solve the inequality $\dfrac{3x+2}{(x+1)(2x)} \le 0$.

9. _____

10. Find all the asymptotes and the intercepts of the function $f(x) = \dfrac{x^2 - 3x + 5}{x + 2}$.

10. _____

FORM A
2 Chapter Test

DATE _____
NAME _____

Directions: Show all work where appropriate. A graphing calculator may be necessary to answer some questions.

1. Divide $x^3 - 2x^2 + 4x - 2$ by $x - 3$.

   1. Quotient: _____
   Remainder: _____

2. What is the remainder when $x^{29} - 7x^{14} + 8$ is divided by $x - 1$?

   2. _____

3. An antique vase is projected to be worth $1,000 in 2 years and $1,300 after 5 years. If the value of the vase continues to appreciate at this same rate, what will it be worth in 8 years?

   3. _____

4. Which one of the following is a polynomial with *real* coefficients that has 2 and $2 - i$ as zeros?

   A. $(x + 2)(x - 2 - i)$
   B. $(x - 2)(x + 2 + i)$
   C. $(x + 2)(x^2 - 4x + 5)$
   D. $(x - 2)(x^2 - 4x + 5)$
   E. $(x + 2)(x^2 + 5)$

   4. _____

5. Find all zeros of $f(x) = x^3 - x^2 + x - 21$ and write a linear factorization of $f(x)$.

   5. Zeros: _____
   $f(x) =$ _____

6. What is the minimum value for the function $y = 2x^2 - 32x + 256$?

   6. _____

7. The line $x = 3$ is the axis of symmetry for the graph of a parabola. If the parabola contains the points $(1, 0)$ and $(4, -3)$, what is the equation for the parabola?

   7. _____

8. A photograph is 4 in. longer than it is wide. If the frame is 2 in. wide, the combined area of the photograph and the frame is 252 in.². Find the dimensions of the photograph without the frame.

   8. _____

9. Graph the function $2x^4 - 3x^3 - 4x^2 + 2x + 2$. Choose a viewing window that shows three local extremum values and all the *x*-intercepts. Make a sketch of the grapher window, and show the viewing window dimensions.

   9.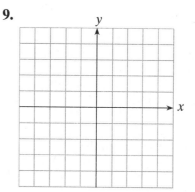

FORM A

2 Chapter Test (continued)   NAME

10. Describe the end behavior of the polynomial function $f(x) = -2x^4 - 3x^3 + 3x - 5$.

10. $f(x) \to$ _____ as $x \to -\infty$;

    $f(x) \to$ _____ as $x \to \infty$

11. Identify the horizontal and vertical asymptotes for the function $f(x) = \dfrac{3x^2}{x^2 - 7x + 12}$.

11. Horizontal: _____

    Vertical: _____

12. Solve the inequality $\dfrac{x - 6}{|2x - 4|} \leq 0$.

12. _____

13. Raymond's distance D from a motion detector is given by the data below. Find a cubic regression equation (with coefficients expressed to the nearest thousandth), and graph it together with a scatter plot of the data.

t(sec)	0.0	0.5	1.0	1.5	2.0	2.5	3.0	3.5	4.0	4.5	5.0
D(m)	2.8	3.9	4.3	4.0	3.3	2.5	1.8	1.2	0.9	1.6	2.7

13.

14. In the space below, identify all asymptotes and intercepts of the function $g(x) = \dfrac{x - 5}{x^2 + x - 6}$. Then sketch a graph of $g(x)$.

14.

15. Tell how the graph of $y = -3 + \dfrac{4}{x + 2}$ can be obtained from the graph of $y = \dfrac{1}{x}$ by using transformations.

15. _____

16. Solve the inequality $\dfrac{(x - 5)^3}{x(x + 2)} \geq 0$.

16. _____

18

FORM B

2 Chapter Test

DATE _____

NAME _____

Directions: Show all work where appropriate. A graphing calculator may be necessary to answer some questions.

1. Divide $x^3 + 3x^2 - 8x + 7$ by $x - 2$.

   1. Quotient: _____

   Remainder: _____

2. What is the remainder when $x^{32} - 5x^{15} + 12$ is divided by $x + 1$?

   2. _____

3. The value of an antique chair is projected to appreciate $60 each year. If the chair will be worth $650 in 2 years, what will it be worth in 10 years?

   3. _____

4. Which one of the following is a polynomial with *real* coefficients that has -2 and $2 + i$ as zeros?

   A. $(x + 2)(x - 2 - i)$
   B. $(x - 2)(x + 2 + i)$
   C. $(x + 2)(x^2 - 4x + 5)$
   D. $(x - 2)(x^2 - 4x + 5)$
   E. $(x + 2)(x^2 + 5)$

   4. _____

5. Find all zeros of $f(x) = x^3 + 7x - 22$ and write a linear factorization of $f(x)$.

   5. Zeros: _____

   $f(x) =$ _____

6. What is the minimum value for the function $y = 3x^2 - 60x + 194$?

   6. _____

7. The line $x = 3$ is the axis of symmetry for the graph of a parabola. If the parabola contains the points $(5, -3)$ and $(-1, 9)$, what is the equation for the parabola?

   7.

8. A swimming pool is 8 ft longer than it is wide. The pool is surrounded by a walkway of width 4 ft. The combined area of the pool and the walkway is 1280 ft². Find the dimensions of the pool without the walkway.

   8. _____

9. Graph the function $y = -3x^4 + 2x^3 + 6x^2 - 5x + 1$. Choose a viewing window that shows three local extremum values and all the *x*-intercepts. Make a sketch of the grapher window, and show the viewing window dimensions.

   9.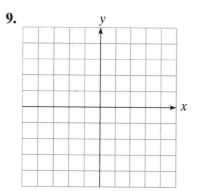

FORM B

2 Chapter Test (continued)   NAME

10. Describe the end behavior of the polynomial function $f(x) = -3x^5 + 2x^4 + 5x - 3$.

10. $f(x) \to$ _____ as $x \to -\infty$;
    $f(x) \to$ _____ as $x \to \infty$

11. Identify the horizontal and vertical asymptotes for the function $f(x) = \dfrac{5x^2}{2x^2 - 11x + 12}$.

11. Horizontal: _____
    Vertical: _____

12. Solve the inequality $(x - 4)\sqrt{x + 2} \geq 0$.

12. _____

13. Jennifer's distance D from a motion detector is given by the data below. Find a cubic regression equation (with coefficients expressed to the nearest thousandth), and graph it together with a scatter plot of the data.

t(sec)	0.0	0.5	1.0	1.5	2.0	2.5	3.0	3.5	4.0	4.5	5.0
D(m)	2.2	1.1	0.7	1.0	1.7	2.5	3.3	4.0	4.4	3.8	2.8

13. _____

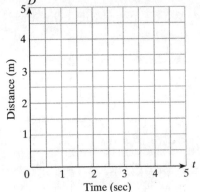

14. In the space below, identify all asymptotes and intercepts of the function $g(x) = \dfrac{x + 6}{x^2 + x - 12}$. Sketch a graph of $g(x)$.

14.

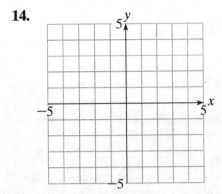

15. Tell how the graph of $y = 5 + \dfrac{2}{x - 4}$ can be obtained from the graph of $y = \dfrac{1}{x}$ by using transformations.

15. _____
_____
_____
_____

16. Solve the inequality $\dfrac{(x - 4)^3}{x(x + 3)} \leq 0$.

16. _____

20

3 Chapter Quiz

3.1–3.4

1. Describe the end behavior of the function $g(x) = \left(\dfrac{2}{3}\right)^x$.

   1. $g(x) \to$ _____ as $x \to -\infty$;

      $g(x) \to$ _____ as $x \to \infty$

2. Find the interval(s) where the function $f(x) = 3^{-x}$ is decreasing.

   A. $(0, \infty)$
   B. $(-\infty, 0)$
   C. $(-\infty, 0) \cup (0, \infty)$
   D. $(-\infty, 3) \cup (3, \infty)$
   E. $(-\infty, \infty)$

   2. _____

3. The population of Powerville is 425,000, and it is increasing at a rate of 4.2% each year. Predict the number of years until the population will be 750,000. Round to two decimal places.

   3. _____

4. Graph the logistic function
   $$f(x) = \dfrac{15}{1 + e^{-2x}}.$$
   Show the viewing window dimensions.

   4.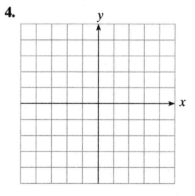

5. Write the exponential equation $3^x = 9$ in logarithmic form.

   5. _____

6. Write the inverse of the function $f(x) = e^{x+2}$.

   6. _____

3.1–3.4

3 Chapter Quiz (continued)   NAME

7. List the transformations used to obtain the graph of $g(x) = 5 \ln(x-1) + 3$ from the graph of $f(x) = \ln x$.

7. _____

8. Which expression is equivalent to $\ln \sqrt[5]{\dfrac{x}{10}}$?

   A. $\dfrac{1}{5} \ln x - \ln 2$

   B. $\dfrac{\ln x}{5 \ln 10}$

   C. $\dfrac{1}{5} \ln x - \dfrac{1}{5} \ln 10$

   D. $\dfrac{\ln 5}{\ln x - \ln 10}$

   E. $\dfrac{x}{10} \ln 5$

8. _____

9. Solve the logarithmic equation $\log_x 5 = 3$ for x. Leave the answer in radical form.

9. _____

10. What is the value of $\log_5 14$, accurate to two decimal places?

10. _____

3.5–3.6

3 Chapter Quiz

1. Which equation is equivalent to $y = 7x^5$?

    A. $\ln y = \ln 7 + 5 \ln x$
    B. $y = \ln 7 + 5 \ln x$
    C. $y = 7 + 5 \ln x$
    D. $\ln y = 7 + 5 \ln x$
    E. $\ln y = \ln 7 - 5 \ln x$

    1. _____

2. Solve the logarithmic equation
   $\log(x + 2) + \log x = 3 \log 2$ algebraically.
   Identify any extraneous solutions.

    2. Solution(s): _____

    Extraneous: _____

3. Determine the domain of the function
   $f(x) = \ln(4x - x^2)$.

    A. $(-\infty, 0) \cup (4, \infty)$
    B. $(0, 4)$
    C. $(-4, 0)$
    D. $(-\infty, 1.39)$
    E. $(-\infty, 0) \cup (0, 4) \cup (4, \infty)$

    3. _____

4. Solve $e^{x^2} + 5x - 1 = 0$ graphically.

    4. _____

5. A casserole is removed from a 375°F oven, and it cools to 200°F after 15 minutes in a 75°F room. How long (from the time it is taken out of the oven) does it take to cool to 100°F? (Hint: Use Newton's Law of Cooling, $T(t) = T_m + (T_0 - T_m)e^{-kt}$.)

    5. _____

6. Find the value of an investment of $5200 invested for 5 years, compounded monthly at the rate of 9.5% APR.

    6. _____

3.5–3.6

3 Chapter Quiz (continued)   NAME

7. A bank advertises that currently a CD pays 6% APR compounded daily (use 365 days a year). What is the effective yield on this CD, to the nearest hundredth?

7. _____

8. Juan contributes $65 monthly into an IRA annuity for 20 years. If the account earns 5.6% annual interest compounded monthly, what is the value of Juan's account after 20 years?

8. _____

9. How long will it take an investment of $750 at 8.75% APR compounded quarterly to grow to $1000?

9. _____

10. A $125,000 mortgage requires monthly payments for 30 years at 7.5% APR. How much is each payment?

10. _____

FORM A

3 Chapter Test

DATE

NAME

Directions: Show all work where appropriate. A graphing calculator may be necessary to answer some questions.

1. Graph the function $y = 0.25^x$.

1.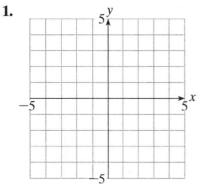

2. (a) Solve for x: $\log_4 64 = x$

   (b) Solve for x: $\log_x 3 = \dfrac{1}{2}$

2. (a) _____

   (b) _____

3. A single-cell amoeba doubles every 4 days. How long would it take one amoeba to produce a population of about 10,000 amoebae?

3. _____

4. What is the natural logarithmic regression equation for the following data? Estimate the y-value for $x = 15$. Express answers to the nearest hundredth.

x	2	4	7	10
y	3	8	11	12

4. Equation: _____

   y-value: _____

5. Show that $f(x) = \dfrac{1}{4} \ln x$ and $g(x) = e^{4x}$ are inverse functions.

5. _____

   _____

6. For the function $f(x) = \dfrac{5}{1 + 4e^{-2x}}$:
   (a) Find the horizontal asymptotes.
   (b) Find the domain and range.
   (c) Describe the end behavior.

6. (a) _____

   (b) _____

   (c) _____

7. Describe the transformations that can be used to transform the graph of
$y = \log_3 x$ to $y = \log_3 (x + 2) + 1$.
Plot the graph of $y = \log_3 (x + 2) + 1$.

7. _____

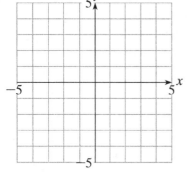

25

FORM A

3 Chapter Test (continued)   NAME

8. Solve the equation $8 - 3 \ln x = 12$. Give an exact answer as well as its decimal approximation (to the nearest hundredth).

8. _____

9. Solve the equation $\log(x - 5) + \log(x - 2) = 1$ algebraically. List any extraneous solutions and explain. If there are no extraneous solutions, explain why not.

9. Solutions: _____

   Extraneous: _____

10. Solve for x: $2^{3x-1} = 16$

10. _____

11. A pie is removed from a 375°F oven and cools to 215°F after 15 minutes in a room at 72°F. How long (from the time it is removed from the oven) will it take the pie to cool to 120°F?

11. _____

12. Let $S = a(1.08)^t$. Solve for t.

12. _____

   A. $\dfrac{\ln S - \ln a}{\ln 1.08}$ B. $\dfrac{\ln(S - a)}{\ln 1.08}$

   C. $\dfrac{\ln S}{\ln a} - \ln 1.08$ D. $\dfrac{\ln S}{\ln a \ln 1.08}$

   E. $\dfrac{\ln(S/a)}{1.08}$

13. When Jose was born, his grandmother deposited $8,000 into an account paying 6% interest, compounded quarterly. How much money will be available to use for college tuition 18 years later?

13. _____

14. A certain company estimates that the computer they plan to buy in 18 months will cost $4,200. How much money should be deposited now into an account paying 5.75% interest, compounded monthly so there will be enough money to pay cash for the computer in 18 months?

14. _____

15. Rosita deposits $250 each month into a retirement account that pays 6.00% APR (0.50% per month). What is the value of this annuity after 20 years?

15. _____

16. To finance their new home, the Colemans have agreed to a $90,000 mortgage loan at 9.25% APR. What will their monthly payments be if the loan has a term of 15 years?

16. _____

   A. $8325.00   B. $11330.60
   C. $6376.63   D. $926.27
   E. $546.25

FORM B

3 Chapter Test

DATE _____

NAME _____

Directions: Show all work where appropriate. A graphing calculator may be necessary to answer some questions.

1. Graph the function $y = 3^{-x}$.

1.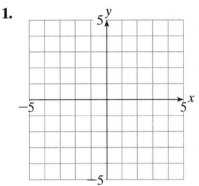

2. (a) Solve for x: $\log_3 81 = x$

   (b) Solve for x: $\log_x 8 = \dfrac{3}{2}$

2. (a) _____

   (b) _____

3. A single-cell amoeba doubles every 3 days. How long would it take one amoeba to produce a population of about 10,000 amoebae?

3. _____

4. What is the natural logarithmic regression equation for the following data? Estimate the y-value for $x = 18$. Express answers to the nearest hundredth.

x	3	5	9	12
y	4	7	10	11

4. Equation: _____

   y-value: _____

5. Show that $f(x) = \dfrac{1}{5} \ln x$ and $g(x) = e^{5x}$ are inverse functions.

5. _____

   _____

6. For the function $f(x) = \dfrac{6}{1 + 2e^{-3x}}$:
   (a) Find the horizontal asymptotes.
   (b) Find the domain and range
   (c) Describe the end behavior.

6. (a) _____

   (b) _____

   (c) _____

7. Describe the transformations that an be used to transform the graph of $y = \log_2 x$ to $y = -\log_2(x + 4)$. Plot the graph of $y = \log_2(x + 4)$.

7. _____

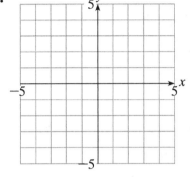

FORM B

3 Chapter Test (continued)   NAME

8. Solve the equation $7 - 4 \log x = 10$. Give an exact answer as well as its decimal approximation (to the nearest hundredth).

    8. _____

9. Solve the equation $\log (x - 4) + \log (x + 5) = 1$ algebraically. List any extraneous solutions and explain. If there are no extraneous solutions, explain why not.

    9. Solutions: _____

       Extraneous: _____

10. Solve for x: $3^{2x-1} = 27$

    10. _____

11. A casserole is removed from an oven at 375°F and cools to 190°F after 25 minutes in a room at 68°F. How long (from the time it is removed from the oven) will it take the casserole to cool to 105°F?

    11. _____

12. Let $S = a(1.06)^t$. Solve for t.

    12. _____

  A. $\dfrac{\ln a - \ln S}{\ln 1.06}$ B. $\dfrac{\ln (1.06)}{\ln a + \ln S}$

  C. $\ln a + \ln 1.06$ D. $\dfrac{\ln (S/a)}{\ln 1.06}$

  E. $\ln \dfrac{(S/a)}{1.06}$

13. When Kara was born, her grandmother deposited $10,500 into an account paying 6.4% interest, compounded quarterly. How much money will be available to use for college tuition 18 years later?

    13. _____

14. A certain company estimates that the computer they plan to buy in 30 months will cost $5,250. How much money should be deposited now into an account paying 6.25% interest, compounded monthly so there will be enough money to pay cash for the computer in 30 months?

    14. _____

15. Juan deposits $200 each month into a retirement account that pays 9.00% APR (0.75% per month). What is the value of this annuity after 30 years?

    15. _____

16. To finance their new home, the Tiballis have agreed to an $80,000 mortgage loan at 8.75% APR. What will their monthly payments be if the loan has a term of 20 years?

    16. _____

  A. $8608.14   B. $7000.00
  C. $706.97   D. $362.50
  E. $4313.29

4.1–4.4

4 Chapter Quiz

1. Point $P(-3, 5)$ is on the terminal side of θ. Evaluate the six trigonometric functions for θ.

   1. $\sin \theta =$ _____ $\cos \theta =$ _____
      $\tan \theta =$ _____ $\csc \theta =$ _____
      $\sec \theta =$ _____ $\cot \theta =$ _____

2. The lengths of the hypotenuse and one leg of a right triangle are 10.5 and 3.7, respectively. Find the measure of the angle opposite the leg whose length is known.

   A. 20.63° **B.** 69.37° **C.** 19.41° **D.** 9.83° **E.** 46.90°

   2. _____

3. Let $\sin x = \dfrac{2}{5}$ and $\dfrac{\pi}{2} < x < \pi$. Find $\tan x$.

   3. _____

4. The wheel of a train engine turns 315° as it moves a distance of 4.75 yards. What is the diameter of the wheel of the engine?

   4. _____

5. If $\sin t = 0.5299$ and $\cos t = 0.8481$, then $\tan t =$ _____.

   5. _____

6. What is the period of $y = \sin 6x$?

   6. _____

7. Find the range of the graph of $y = 6 \sin \pi x$.

   A. $[-\pi, \pi]$ **B.** $[-6, 6]$ **C.** $[-2\pi, 2\pi]$
   D. $[-3, 3]$ **E.** $(-\infty, \infty)$

   7. _____

8. Find the amplitude of the graph of $y = 3 + 2\cos(x - \pi)$.

   8. _____

9. A guy wire is stretched from the top of a 200-ft transmission tower to a point on the ground 130 ft from the center of the base of the tower. What angle, in degrees, does the guy wire make with the ground?

   9. _____

10. Sketch the graph of $y = 2 + 4 \sin \dfrac{2}{3}x$.

    10.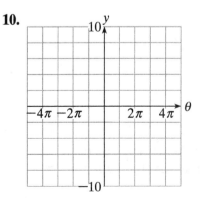

29

4 Chapter Quiz

4.5–4.8

DATE

NAME

1. Find the period of $y = 3 \tan(2x - 1)$.

   1. _____

2. Find the domain of $f(x) = \sin^{-1}(5x)$.

   2. _____

   A. $[-5, 5]$

   B. $\left[-\dfrac{1}{5}, \dfrac{1}{5}\right]$

   C. $[-1, 1]$

   D. $\left[-\dfrac{\pi}{2}, \dfrac{\pi}{2}\right]$

   E. $\left[-\dfrac{\pi}{10}, \dfrac{\pi}{10}\right]$

3. Sketch one period of the graph of $y = \sin 2x + \sin 3x$.

   3.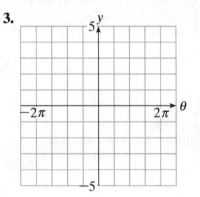

4. State the domain, range, and period of the function $y = |\sin^3 x|$.

   4. _____

5. Sketch the graph of $y = 3 \sec x$ for $-\pi \leq x \leq \pi$.

   5.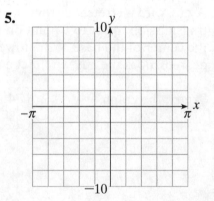

6. Estimate the value of the amplitude graphically of the sinusoidal function $f(x) = 3 \sin 3x + 2 \cos 3x$.

   6. _____

4 Chapter Quiz (continued) NAME

4.5–4.8

7. Find an algebraic expression equivalent to $\sec(\tan^{-1} x)$.   7. _____

   A. $\sqrt{x^2 + 1}$

   B. $\dfrac{1}{\sqrt{x^2 + 1}}$

   C. $\dfrac{\sqrt{x^2 + 1}}{x}$

   D. $\sqrt{x^2 - 1}$

   E. $x^2 + 1$

8. From a 75-foot observation tower, a controller sights a hot-air balloon at an angle of 20° elevation. If the horizontal distance from the tower to the balloon is 300 feet, how much higher in the air is the balloon?   8. _____

9. Jeana sights the top of a building and finds the angle of elevation to be 35°. She moves 100 feet closer and finds that the angle is now 40°. What is the height of the building?   9. _____

10. A certain company that produces a product with seasonal demands projects that monthly sales for the next two years can be modeled by   10. _____

    $y = 14.25 + 0.25x + 2.8 \sin\left(\dfrac{\pi x}{6}\right)$. In the model, y is the sales in thousands of units and x is the time in months, with $t = 1$ representing January, 2006. According to this model, what will be the projected sales for July 2006?

FORM A

4 Chapter Test

DATE _____

NAME _____

Directions: Show all work where appropriate. A graphing calculator may be necessary to answer some questions.

1. Evaluate all six trigonometric functions of the angle θ.

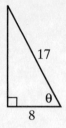

1. $\sin \theta =$ _____ $\cos \theta =$ _____

   $\tan \theta =$ _____ $\csc \theta =$ _____

   $\sec \theta =$ _____ $\cot \theta =$ _____

2. Compute sec 2.

2. $\sec 2 =$ _____

3. The wheels on a truck are turning at 630 revolutions per minute as the truck travels at 60 miles per hour. What is the radius, in inches, of the wheels on the truck?

3. _____

4. Solve the right triangle $\triangle ABC$ for all its unknown parts if $\beta = 38°$ and $b = 4.5$.

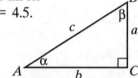

4. $\alpha =$ _____

   $a =$ _____

   $c =$ _____

5. Which transformation was *not* performed on $y = \sin x$ to obtain $y = -2 \sin \left(3x + \dfrac{\pi}{3} \right)$?

   A. Horizontal shift left $\dfrac{\pi}{9}$ units
   B. Horizontal stretch by a factor of 3
   C. Vertical stretch by a factor of 2
   D. Reflection through the x-axis
   E. Horizontal shrink by a factor of $\dfrac{1}{3}$

5. _____

6. State the amplitude, period, phase shift, and vertical translation of the sinusoid.

   $y = 2 + 6 \sin \left(3x - \dfrac{\pi}{4} \right)$

6. Amplitude: _____

   Period: _____

   Phase shift: _____

   Vert. trans.: _____

7. For each of the following functions, determine if the graph shows damped oscillation. If so, state the damping factor.

   (a) $f(x) = x + \sin x$ (b) $f(x) = 3^{-x} \cos 4x$
   (c) $f(x) = -4x \cos 2x$ (d) $f(x) = 2 \sin 3x$
   (e) $F(x) = 0.35 e^{-0.07x} \cos 2x$

7. (a) _____
   (b) _____
   (c) _____
   (d) _____
   (e) _____

FORM A
4 Chapter Test (continued)   NAME

8. In the space below, explain how the graph of the function $y = 3 \csc \frac{1}{2}x$ is related to a basic trigonometric graph. Determine the period, domain, range, zeros, and asymptotes (if any). Then sketch the graph of the function.

8.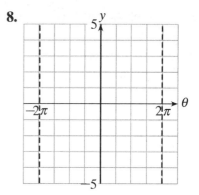

9. Solve the equation $\cot x = 2.7$ for x in the interval $\pi \leq x \leq 3\pi$ graphically. Estimate answers to the nearest hundredth.

9. _____

10. Find approximate values of a, b, and h (to the nearest hundredth) so that $f(x) \approx a \sin[b(x - h)] + k$.
    $f(x) = 2 \sin 5x - 4 \cos 5x$

10. $a \approx$ _____
    $b \approx$ _____
    $h \approx$ _____

11. Which is the correct value of $\sin^{-1}(0.6)$? (Express the answer in radians to the nearest hundredth.)
    A. 1.77   **B.** 0.64   **C.** 95.49   **D.** 36.87   **E.** 0.03

11. _____

12. The angle of elevation from the top of a building to the top of a nearby taller building is 56.5° and the angle of depression to the bottom of the taller building is 15.3°. If the short building is 50.2 m high, what is the height of the tall building?

12. _____

13. Find an equivalent *algebraic* expression not involving trig functions for $\cos(\sin^{-1} u)$. Explain why your expression includes (or does not include) a $\pm$ sign.

13. _____

14. For the next 10 years, a small company's business volume can be modeled by the function
    $f(x) = 80(1.03)^x + 3 \sin \frac{\pi x}{4}$, where x is the number of years after 2006 and f is the sales in millions of dollars.
    (a) What are the company's sales in 2006?
    (b) What does the model predict for sales in 2012?
    (c) How many years are in each economic cycle for this company?

14. (a) _____
    (b) _____
    (c) _____

15. Find the period and graph the function for $-2\pi \leq x \leq 2\pi$. Be sure to label your graph with the viewing window dimensions.
    $y = 4 \cos 2x + 6 \sin(3x - 2)$

15. Period: _____

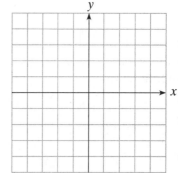

33

FORM B

DATE _____

4 Chapter Test

NAME _____

Directions: Show all work where appropriate. A graphing calculator may be necessary to answer some questions.

1. Evaluate all six trigonometric functions of the angle θ.

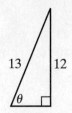

1. $\sin \theta = $ _____ $\cos \theta = $ _____

   $\tan \theta = $ _____ $\csc \theta = $ _____

   $\sec \theta = $ _____ $\cot \theta = $ _____

2. Compute csc 3.

2. csc 3 = _____

3. The wheels on a truck are turning at 550 revolutions per minute as the truck travels at 55 miles per hour. What is the radius, in inches, of the wheels on the truck?

3. _____

4. Solve the right triangle $\triangle ABC$ for all its unknown parts if $\alpha = 43°$ and $a = 3.5$.

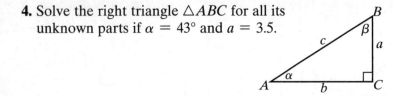

4. $\beta = $ _____

   $b = $ _____

   $c = $ _____

5. Which transformation was *not* performed on $y = \cos x$ to obtain $y = -4 \cos\left(2x + \dfrac{\pi}{3}\right)$?

5. _____

   A. Vertical stretch by a factor of 4
   B. Reflection through the x-axis
   C. Horizontal stretch by a factor of 2
   D. Horizontal shrink by a factor of $\dfrac{1}{2}$
   E. Horizontal shift left $\dfrac{\pi}{6}$ units

6. State the amplitude, period, phase shift, and vertical translation of the sinusoid.

   $y = -3 + 2\sin\left(4x - \dfrac{\pi}{5}\right)$

6. Amplitude: _____

   Period: _____

   Phase shift: _____

   Vert. trans.: _____

7. For each of the following functions, determine if the graph shows damped oscillation. If so, state the damping factor.

   (a) $f(x) = 4^{-x} \cos 3x$ (b) $f(x) = x + \cos x$

   (c) $f(x) = -5x \cos 3x$ (d) $f(x) = 4 \sin \dfrac{x}{3}$

   (e) $F(x) = 0.45 e^{-0.1x} \cos 2x$

7. (a) _____

   (b) _____

   (c) _____

   (d) _____

   (e) _____

FORM B

4 Chapter Test (continued) NAME

8. In the space below, explain how the graph of the function $y = -\sec 2x$ is related to a basic trigonometric graph. Determine the period, domain, range, zeros, and asymptotes (if any). Then sketch the graph of the function.

8.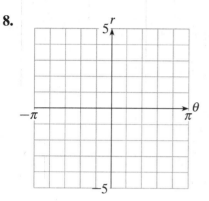

9. Solve the equation $\cot x = -3.2$ for x in the interval $\pi \leq x \leq 3\pi$ graphically. Estimate answers to the nearest hundredth.

9. _____

10. Find approximate values of a, b, and h (to the nearest hundredth) so that $f(x) \approx a \sin [b(x - h)]$.
$f(x) = 4 \sin 3x - 6 \cos 3x$

10. $a \approx$ _____
$b \approx$ _____
$h \approx$ _____

11. Which is the correct value of $\cos^{-1}(0.4)$. (Express the answer in radians to the nearest hundredth.)

   A. 1.00   B. 1.09   C. 1.53   D. 87.71   E. 1.16

11. _____

12. The angle of elevation from the top of a building to the top of a nearby taller building is 48.3° and the angle of depression to the bottom of the taller building is 18.1°. If the short building is 61.7 m high, what is the height of the tall building?

12. _____

13. Find an equivalent *algebraic* expression not involving trig functions for $\csc (\cos^{-1} u)$. Explain why your expression includes (or does not include) a $\pm$ sign.

13. _____

14. For the next ten years, a small company's business volume can be modeled by the function
$f(x) = 68(1.07)^x + 5 \sin \dfrac{\pi x}{3}$, where x is the number of years after 2006 and f is the sales in millions of dollars.
   (a) What are the company's sales in 2006?
   (b) What does the model predict for sales is 2015?
   (c) How many years are in each economic cycle for this company?

14. (a) _____
    (b) _____
    (c) _____

15. Find the period and graph the function for $-2\pi \leq x \leq 2\pi$. Be sure to label your graph with the viewing window dimensions.
$y = 7 \cos 3x - 3 \sin (2x - 5)$

15. Period: _____

5.1–5.3

5 Chapter Quiz

DATE

NAME

1. Simplify the expression $\dfrac{\csc \theta}{\sin^2 \theta + \sec \theta + \cos^2 \theta - 1}$.

1. _____

2. Factor the expression $\dfrac{\cot^2 x}{\csc x - 1}$ and use fundamental identities to simplify.

2. _____

3. Use the grapher to determine which of the following is an identity.

   A. $\sin^2 x - \cos x = 1 + 2 \sin x$
   B. $\sin^3 x - \sin x = -\cos^2 x \sin x$
   C. $\sin(x + \pi) = \sin x$
   D. $\tan 2x = 2 \sin x \cos x$
   E. $\cos 2x + \sin 2x = 1$

3. _____

4. Prove the identity: $\dfrac{\sin \theta}{1 + \cos \theta} + \dfrac{1 + \cos \theta}{\sin \theta} = 2 \csc \theta$

4. _____
_____
_____
_____
_____
_____

5. Use the sum identities to prove the identity:
$\cos \theta + \cos 2\theta + \cos 3\theta = (2 \cos \theta + 1) \cos 2\theta$

5. _____
_____
_____
_____
_____
_____

5.1–5.3

5 Chapter Quiz (continued)   NAME

6. Which of the following is correct?   6. _____

   A. $\sin(-x) = \sin x$, $\cos(-x) = \cos x$,
      $\tan(-x) = -\tan x$

   B. $\sin(-x) = -\sin x$, $\cos(-x) = -\cos x$,
      $\tan(-x) = \tan x$

   C. $\sin(-x) = \sin x$, $\cos(-x) = -\cos x$,
      $\tan(-x) = -\tan x$

   D. $\sin(-x) = -\sin x$, $\cos(-x) = \cos x$,
      $\tan(-x) = -\tan x$

   E. $\sin(-x) = -\sin x$, $\cos(-x) = \cos x$,
      $\tan(-x) = \tan x$

7. Solve the equation $2 \sin^2 x \cos x = \cos x$ by factoring   7. _____
   and/or extracting square roots.

8. Find all solutions of the equation   8. _____
   $2 \sin^2 x + 3 \sin x - 2 = 0$ on the interval $[0, 2\pi)$.
   (Your answer should be exact.)

9. Solve the inequality $|\cot x| \leq \sqrt{3}$ for $0 \leq x \leq 2\pi$.   9. _____

10. Confirm or disprove the following: $2 \cos x = \sin 2x$.   10. _____
    If this is an identity, provide a confirmation. If it is not
    an identity, provide a counter example.

5.4–5.6

5 Chapter Quiz

1. In triangle ABC, $\alpha = 42°$, $\beta = 25°$, and $b = 6$. Find the remaining sides and angle.

   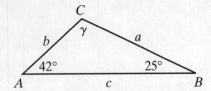

   1. _____

2. For positive values of a, c, and γ, where $\gamma < 180°$ and $c < a \sin \gamma$, how many triangles can be formed?

   A. one unique triangle
   B. two triangles are possible
   C. no triangles are possible
   D. more information is necessary to decide.

   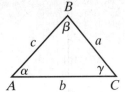

   2. _____

3. A civil engineer needs to measure the length of a proposed tunnel through a mountain. The engineer chooses a point that can be seen from each end of the tunnel and measures the distance to each end of the tunnel to be 3,000 meters and 4,200 meters, respectively. If the angle between the two distances is 113°, what is the length of the tunnel?

   3. _____

4. What is the area of $\triangle ABC$ if $a = 8$, $b = 10$, and $c = 15$?

   4. _____

5. What is the area of $\triangle ABC$ if $\alpha = 42°$, $b = 17$, and $c = 25$?

   5. _____

6. The angle of elevation to the bottom of a steeple of a church is 35.2° and the angle of elevation to the top of the steeple is 48.7°. If the angles of elevation are measured from a point on the ground 60 meters from the church steeple, what is the height of the steeple?

   6. _____

38

5.4–5.6

5 Chapter Quiz (continued) NAME

7. Which of the following is not a valid completion of the double-angle identity for $\cos 2\theta$?

$\cos 2\theta =$

A. $2\cos^2 a - 1$

B. $1 - 2\sin^2 \theta$

C. $\sin^2 \theta - \cos^2 \theta$

D. $\cos^2 \theta - \sin^2 \theta$

7. _____

8. Use a power-reduction formula to find the exact value of $\sin\left(\dfrac{\pi}{8}\right)$. (Hint: What is $\sin^2\left(\dfrac{\pi}{8}\right) = ?$)

8. _____

9. Two points are 200 ft apart on opposite sides of a tree. The angles of elevation from each point to the top of the tree are 30° and 35°. What is the height of the tree?

9. _____

10. Prove the identity:

$$\dfrac{\sin^3 A - \cos^3 A}{\sin^2 A - \cos^2 A} = \dfrac{2 + \sin 2A}{2(\sin A + \cos A)}$$

10. _____
_____
_____
_____

FORM A
5 Chapter Test

Directions: Show all work where appropriate. A graphing calculator may be necessary to answer some questions.

1. Use the fundamental identities to change the expression to one involving only sines and cosines. Then simplify. Show all your steps.
$$\sec^2 x - \frac{\sec^2 x}{\csc^2 x}$$

2. Prove the identity: $(\tan x + 1)^2 = \dfrac{1 + 2 \sin x \cos x}{\cos^2 x}$

3. What is the general solution to $\sin x \tan x - \sin x + \tan x - 1 = 0$?

4. Use a grapher to conjecture whether the equation is likely to be an identity. If it is not, give a counter example.
$$\sin^2 x \cos^4 x = \cos^2 x + \cos^4 x - \cos^6 x$$

5. Use a sum or difference identity to find the exact value of $\sin 165°$.

6. Prove the identity: $\cos 3x = \cos^3 x - 3 \sin^2 x \cos x$

7. Use the half-angle identities to express $\sin 3C$ in terms of a function of $6C$.

In Problems 8–9, consider the triangle ABC. Express your answers as decimals rounded to the nearest hundredth. If more than one triangle is possible, give all possible answers.

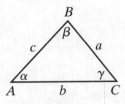

8. If $a = 9$, $b = 8$, and $\alpha = 75°$, find β.

9. If $a = 8$, $b = 10$, and $\alpha = 67°$, how many triangles are determined?

FORM A

5 Chapter Test (continued)   NAME

10. A golf ball is hit with initial velocity v_o (in feet per second) and trajectory angle θ. The horizontal distance traveled before it bounces is given by $d = \dfrac{v_o^2}{16} \sin \theta \cos \theta$. Find the angle of trajectory needed to hit a green 700 feet away on the first bounce if the initial velocity is 170 ft/sec.

10. _____

11. The lengths of the sides of a triangular parcel of land are 300 ft, 350 ft, and 450 ft. What is the area, in square feet, of the parcel of land?

11. _____

12. Due to the wind, a tree grew so that it leaned 8° from the vertical. From a point on the ground 28 meters from the base of the tree, the angle of elevation to the top of the tree is 24.6°. What is the height of the tree?

12. _____

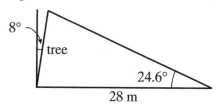

13. Kelly must find the distance between points A and C on opposite sides of a lake. She locates a point B that is 425 ft from A and 672 ft from C. If the angle at B is 68°, what is the distance AC?

13. _____

14. The angles of elevation to the top of a tower from two points that are on the same side of the tower and 30 meters apart are 37.6° and 46.3°. What is the height of the tower?

14. _____

15. Suppose the angle θ is formed by two sides of length 8 in an isosceles triangle. Find an expression for the area $A(\theta)$ of the triangle. Then find the value(s) of θ so that the triangle's area is 25.

15. $A(\theta) =$ _____

$\theta \approx$ _____

Chapter 5 Test — Form B

Directions: Show all work where appropriate. A graphing calculator may be necessary to answer some questions.

1. Use the fundamental identities to change the expression to one involving only sines and cosines. Then simplify. Show all your steps.
$$\tan^2 x - \frac{\csc^2 x}{\cot^2 x}$$

2. Prove the identity: $(1 + \cot x)^2 = \dfrac{1 + 2 \sin x \cos x}{\sin^2 x}$

3. What is the general solution to $\cos^2 x \sin x - \sin x - \cos^2 x + 1 = 0$?

4. Use a grapher to conjecture whether the equation is likely to be an identity. If it is not, give a counter example.
$$\sin^4 x \cos^2 x = \cos^2 x + \sin^4 x - \cos^6 x$$

5. Use a sum or difference identity to find the exact value of $\cos 195°$.

6. Prove the identity:
$$\sin 3x = 4 \sin x \cos^2 x - \sin x$$

7. Use the half-angle identities to express $\cos 4C$ in terms of a function of $8C$.

In Problems 8–9, consider the triangle ABC. Express your answers as decimals rounded to the nearest hundredth. If more than one triangle is possible, give all possible answers.

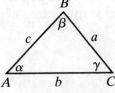

8. If $a = 9$, $b = 8$, and $\alpha = 65°$, find β.

9. If $a = 10$, $b = 8$, and $\alpha = 67°$, how many triangles are determined?

FORM B

5 Chapter Test (continued)   NAME

10. A golf ball is hit with initial velocity v_o (in feet per second) and trajectory angle θ. The horizontal distance traveled before it bounces is given by $d = \dfrac{v_o^2}{16} \sin\theta \cos\theta$. Find the angle of trajectory needed to hit a green 800 feet away on the first bounce if the initial velocity is 200 ft/sec.

10. _____

11. The lengths of the sides of a triangular parcel of land are 200 ft, 250 ft, and 350 ft. What is the area, in square feet, of the parcel of land?

11. _____

12. Due to the wind, a tree grew so that it leaned 9° from the vertical. From a point on the ground 30 meters from the base of the tree, the angle of elevation to the top of the tree is 26.3°. What is the height of the tree?

12. _____

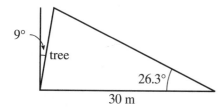

13. Randy must find the distance between points B and C on opposite sides of a lake. He locates a point A that is 385 ft from B and 546 ft from C. If the angle at A is 81°, what is the distance BC?

13. _____

14. The angles of elevation to the top of a tower from two points that are on the same side of the tower and 30 meters apart are 39.2° and 42.6°. What is the height of the tower?

14. _____

15. Suppose the angle θ is formed by two sides of length 10 in an isosceles triangle. Find an expression for the area $A(\theta)$ of the triangle. Then find the value(s) of θ so that the triangle's area is 37.

15. $A(\theta) =$ _____

$\theta \approx$ _____

43

P-5 Midterm Exam A

Show all work where appropriate. A graphing calculator may be necessary to answer some questions.

1. Solve the double inequality $-12 < \dfrac{5x - 6}{2} \leq -7$.

   Express your answer in interval notation.

   1. _____

2. Show that the points $A(1, -3)$, $B(-8, 1)$, $C(-4, 10)$, and $D(5, 6)$ are the vertices of a square.

   2. _____

3. Rewrite the expression $\dfrac{\sqrt[7]{x^9}}{\sqrt[5]{x^6}}$ using a single radical.

   3. _____

4. What is the equation, in slope-intercept form, of the line that contains the points $(-2, 7)$ and $(3, 5)$?

   4. _____

5. The table shows average salaries for employees of Mediocre Tools, Inc. Find a linear regression equation for the data and graph the equation with a scatter plot of the data. (Use $x = 0$ for 1975, and express values to the nearest hundredth.) Use this data to estimate the average salary in 2008.

   5. _____

   2008: _____

   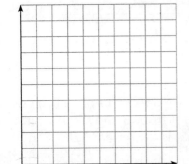

Year	Average Salary ($)
1975	15,642
1980	16,580
1985	17,409
1990	19,150
1995	20,491
2000	21,925

6. Determine the range of $f(x) = 13 - 20x - x^2 - 3x^4$. Express your answer in interval notation.

   6. _____

7. Solve the equation $\log(x - 6) + \log(x - 3) = 1$. List any extraneous roots and explain.

   7. Solution(s): _____

   Ext. root(s): _____

8. Solve the equation graphically. $|4x - 3| = 5\sqrt{x + 4}$

   8. _____

9. Let $f(x) = \sqrt{x - 3}$ and $g(x) = x^2 + 1$. Compute $g \circ f$ and state its domain in interval notation.

   9. _____

P-5 Midterm Exam A (continued)

10. Three sides of a fence and a wall form a rectangular enclosure. The total length of a fence used for the three sides is 240 feet. Let x be the length of the two sides perpendicular to the wall as shown.

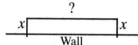

Write the area A of the enclosure as a function of the length x. Then find the value(s) of x for which the area is 5500 ft².

10. _____

11. Solve the inequality $|x - 8| > 4$. Express your answer in interval notation.

11. _____

12. The graph of the function $x^2 + y^2 + 20x - 16y + 80 = 0$ is a circle. What are the coordinates of the center and the length of the radius?

12. Center: _____
Radius: _____

13. Find the equation of the parabola that passes through $(4, 7)$ and has vertex $(2, -5)$. Express your answer in the standard form for a quadratic function.

13. _____

14. Graph the relation defined by the parametric equations when $-2 \leq t \leq 3$.

$x(t) = t^2 - 5, y(t) = t - 1$.

14.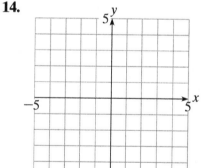

15. The following three transformations are applied (in the given order) to the graph of $y = x^2$:

I. A vertical stretch by a factor of 3
II. A horizontal shift right 5 units
III. A vertical shift down 6 units

Write an equation for the graph produced as a result of applying these transformations.

15. _____

16. Which of the following could represent a complete graph of $f(x) = ax - x^3$ where a is a real number?

16. _____

A.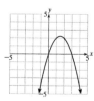
B.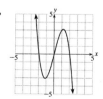
C.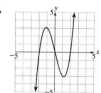
D.

P-5 Midterm Exam A (continued)

17. For the function $f(x) = \dfrac{2(x^2 - 16)}{x^2 - 9}$, what are the intercepts and what are the asymptotes?

17. x-intercept: _____

y-intercept: _____

Vert. asym.: _____

Hor. asym.: _____

18. Use synthetic division to show that the number $c = -2$ is a lower bound for the zeros of $f(x) = x^4 - 3x^3 - 4x^2 + 8x - 2$. Be sure to reference any properties, theorems or tests used.

18. _____

19. Find two complex number roots for the equation $x^2 - 4x + 21 = 0$.

19. _____

20. Find a polynomial of degree 3 with real number coefficients that has $3 + i$ and -7 as zeros.

20. _____

21. For the function $y = 2 \ln(x + 3)$, what is the inverse function, f^{-1}?

21. _____

22. The graph of $y = 2 - a^{x+3}$ for $a > 1$ is best represented by which graph?

22. _____

A.  B. 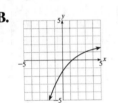 C. 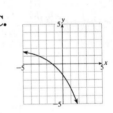 D.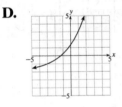

23. Describe transformations that can be used to transform the graph of $\log x$ to a graph of $f(x) = 4 \log(x + 2) - 3$.

23. _____

24. Geoff invests $5,000 at 8.2% for 12 years. If the interest is compounded daily (use 360 days for the year), what will the investment be worth in 12 years?

24. _____

25. Arturo invests $2700 in a savings account that pays 9% interest, compounded quarterly. If there are no other transactions, when will his balance reach $4550?

25. _____

26. In a certain state park, the number of elk present after t years is modeled by the function $P(t) = \dfrac{1216}{1 + 75e^{-0.03t}}$.

(a) What was the initial population of elk?

(b) When will the number of elk be 750?

(c) What is the maximum number of elk possible in the park?

26. (a) _____

(b) _____

(c) _____

P-5 Midterm Exam A (continued)

27. Evaluate all six trigonometric functions of the angle θ for the triangle given below.

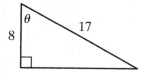

27. $\sin \theta =$ _____

$\tan \theta =$ _____

$\csc \theta =$ _____

$\cos \theta =$ _____

$\sec \theta =$ _____

$\cot \theta =$ _____

28. Solve the right triangle $\triangle ABC$ for all its unknown parts if $\beta = 38°$ and $b = 4.5$.

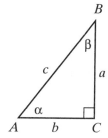

28. $\alpha =$ _____

$a =$ _____

$c =$ _____

29. Which transformation was *not* performed on $y = \sin x$ to obtain $y = -2\sin(3x + \pi/3)$?

29. _____

   A. Horizontal shift left by $\pi/9$ units
   B. Horizontal stretch by a factor of 3
   C. Vertical stretch by a factor of 2
   D. Reflection through the x-axis

30. At the top of a radio signal tower there is an antenna with a light on the end of it. From a point on the ground 500 feet from the base of the tower, the angle of elevation to the tip of the light is 35.6° and the angle of elevation to the bottom of the antenna is 30.4°. What is the height of the antenna, including the light?

30. _____

31. Simplify: 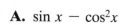 $(\csc x - \tan x)\sin x \cos x =$

31. _____

   A. $\sin x - \cos^2 x$ **B.** $\cos x - \sin^2 x$
   C. $\sin^2 x + \cos x$ **D.** $\cos^2 x - \sin x$

32. For the next eight years, a small company's business volume can be modeled by the function

$$f(x) = 108(1.02)^x + 2\sin\frac{\pi x}{3},$$ where x is the number of years after 2006 and f is the sales in millions of dollars.

32. (a) _____

(b) _____

(c) _____

   (a) What are the company's sales in 2006?
   (b) What does the model project for sales in 2011?
   (c) How many years are in each economic cycle for this company?

P-5 Midterm Exam B

Show all work where appropriate. A graphing calculator may be necessary to answer some questions.

1. Solve the double inequality $-8 \leq \dfrac{5x+4}{2} < 7$.

   Express your answer in interval notation.

   1. _____

2. Show that the points $A(4, 2)$, $B(-1, 0)$, $C(-3, 5)$, and $D(2, 7)$ are the vertices of a square.

   2. _____
   _____
   _____
   _____
   _____

3. Rewrite the expression $\dfrac{\sqrt[5]{x^7}}{\sqrt[3]{x^4}}$ using a single radical.

   3. _____

4. What is the equation, in slope-intercept form, of the line that passes through the points $(-3, 4)$ and $(5, 7)$?

   4. _____

5. The table shows average salaries for employees of Fairto Middlin Corporation. Find a linear regression equation for the data and graph the equation with a scatter plot of the data. (Use $x = 0$ for 1975, and express values to the nearest hundredth.) Use this data to estimate the average salary in 2012.

   5. _____

   2008: _____

   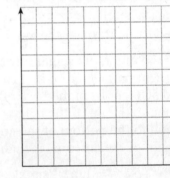

Year	Average Salary ($)
1975	18,632
1980	19,563
1985	21,324
1990	22,817
1995	25,099
2000	26,103

6. Determine the range of $f(x) = 3 + 15x - x^3 - 2x^4$. Express your answer in interval notation.

   6. _____

7. Solve the equation $\log(x - 4) + \log(x + 5) = 1$. List any extraneous roots and explain.

   7. Solution(s): _____

   Ext. root(s): _____
   _____
   _____

8. Solve the equation graphically. $|5x - 4| = (\sqrt{x + 3})5$

   8. _____

9. Let $f(x) = x^2 - 1$ and $g(x) = \sqrt{x + 5}$. Compute $f \circ g$ and state its domain in interval notation.

   9. _____

P-5 Midterm Exam B (continued)

10. Three sides of a fence and a wall form a rectangular enclosure. The total length of a fence used for the three sides is 160 feet. Let x be the length of the two sides perpendicular to the wall as shown.

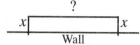

Write the area A of the enclosure as a function of the length x. For what values of x will the area be 5,500 square feet?

10. _____

11. Solve the inequality $|x - 5| \geq 3$. Express your answer in interval notation.

11. _____

12. The graph of the function $x^2 + y^2 - 24x + 12y + 100 = 0$ is a circle. What are the coordinates of the center and the length of the radius?

12. Center: _____
    Radius: _____

13. Find the equation of the parabola that passes through $(-6, 22)$ and has vertex $(2, -5)$. Express your answer in the standard form for a quadratic function.

13. _____

14. Graph the relation defined by the parametric equations when $-3 \leq t \leq 2$.

$x(t) = 5 - t^2, y(t) = t + 2.$

14.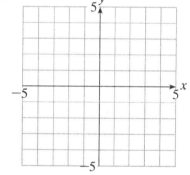

15. The following three transformations are applied (in the given order) to the graph of $y = x^2$:

   I. A vertical stretch by a factor of 0.3
   II. A horizontal shift left 5 units
   III. A vertical shift up 6 units

Write an equation for the graph produced as a result of applying these transformations.

15. _____

16. Which of the following could represent a complete graph of $f(x) = ax + x^3$ where a is a real number?

16. _____

A. 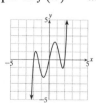 B. 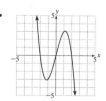 C. 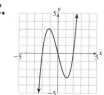 D.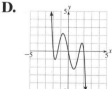

P-5 Midterm Exam B (continued)

17. For the function $f(x) = \dfrac{3(x^2 - 25)}{x^2 - 4}$, what are the intercepts and what are the asymptotes?

17. x-intercept _____
    y-intercept _____
    Vert. asym. _____
    Hor. asym. _____

18. Use synthetic division to show that the number $c = -3$ is a lower bound for the zeros of $f(x) = x^4 - 2x^3 - 8x^2 + 3x - 10$. Be sure to reference any properties, theorems or tests used.

18. _____

19. Find two complex number roots for the equation $x^2 - 6x + 58 = 0$.

19. _____

20. Find a polynomial of degree 3 with real number coefficients that has $8 + 2i$ and 5 as zeros.

20. _____

21. For the function $y = 4 + \ln(x + 1)$, what is the inverse function, f^{-1}?

21. _____

22. The graph of $y = -3 + a^{2-x}$ for $a > 1$ is best represented by which graph?

22. _____

A. 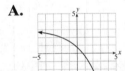 B. 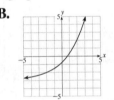 C. 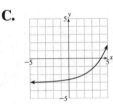 D.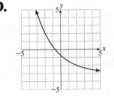

23. Describe transformations that can be used to transform the graph of $y = \ln x$ to a graph of $5 - \ln(x - 4)$.

23. _____

24. Martha invests $6,000 at 7.5% for 8 years. If the interest is compounded daily (use 360 days for the year), what will the investment be worth in 8 years?

24. _____

25. Melanie invests $3,500 in a savings account that pays 8% interest, compounded monthly. If there are no other transactions, when will her balance reach $4,300?

25. _____

26. In a certain national forest, the number of moose present after t years is modeled by the function
$$P(t) = \dfrac{1518}{1 + 65e^{-0.05t}}.$$

(a) What was the initial population of moose?
(b) When will the number of moose be 850?
(c) What is the maximum number of moose possible in the forest?

26. (a) _____
    (b) _____
    (c) _____

P-5 Midterm Exam B (continued)

27. Evaluate all six trigonometric functions of the angle θ for the triangle given below.

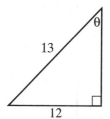

27. $\sin \theta =$ _____

$\tan \theta =$ _____

$\csc \theta =$ _____

$\cos \theta =$ _____

$\sec \theta =$ _____

$\cot \theta =$ _____

28. Solve the right triangle $\triangle ABC$ for all its unknown parts if $\alpha = 43°$ and $a = 3.5$.

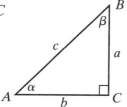

28. $\alpha =$ _____

$a =$ _____

$c =$ _____

29. Which transformation was *not* performed on $y = \cos x$ to obtain $y = -4\cos(2x + \pi/3)$?

29. _____

A. Vertical stretch by a factor of 4

B. Horizontal stretch by a factor of 2

C. Horizontal shift left by $\pi/6$ units

D. Reflection through the x-axis.

30. At the top a radio signal tower there is an antenna with a light on the end of it. From a point on the ground 800 feet from the base of the tower, the angle of elevation to the tip of the light is 48.8° and the angle of elevation to the bottom of the antenna is 42.7°. What is the height of the antenna, including the light?

30. _____

31. Simplify: $(\sec x - \csc x)\sin x \cos x =$

31. _____

A. $\sin x - \cos x$ B. $\cos x - \sin x$

C. $\sin x + \cos x$ D. $\cos x + \sin x$

32. For the next 15 years, a small company's business volume can be modeled by the function $f(x) = 47(1.09)^x + 3\sin\dfrac{\pi x}{6}$, where x is the number of years after 2006 and f is the sales in millions of dollars.

32. (a) _____

(b) _____

(c) _____

(a) What are the company's sales in 2006?

(b) What does the model project for sales in 2013?

(c) How many years are in each economic cycle for this company?

6.1–6.3

6 Chapter Quiz

DATE _____

NAME _____

Show your work on the following questions. A graphing calculator may be necessary to answer some problems. Answers should be accurate to two decimal places.

1. What is the unit vector in the direction of $\mathbf{v} = \langle 6, -8 \rangle$?

   1. _____

2. Which of the following vectors are orthogonal?

   A. $\langle 1, 2 \rangle$ and $\langle 2, 4 \rangle$
   B. $\langle 1, 2 \rangle$ and $\langle -1, -2 \rangle$
   C. $\langle -1, -2 \rangle$ and $\langle 1, -1 \rangle$
   D. $\langle -1, -2 \rangle$ and $\langle -2, 1 \rangle$

   2. _____

3. A plane is flying on a bearing of 35° east of north at 550 mph. Express the velocity of the plane as a vector. Assume that there is no wind.

   3. _____

4. Find the angle between $\mathbf{u} = \langle -2, 5 \rangle$ and $\mathbf{v} = \langle 1, -3 \rangle$.

   4. _____

5. Find the components of the vector $\mathbf{v}$ with direction angle 242° and length 5.

   5. _____

6 Chapter Quiz (continued) NAME

6. Eliminate the parameter from the curve C defined by the parametric equations $x(t) = t^2 + 2$ and $y(t) = 5 - t$.

6. _____

7. What is the parameterization of the line through the points $(5, -4)$ and $(2, 6)$?

7. _____

8. What is the vector projection of $\mathbf{u} = \langle 8, 2 \rangle$ onto $\mathbf{v} = \langle 4, -4 \rangle$?

8. _____

9. What is the work done in lifting a 45-pound child 5 feet off the ground?

   A. 9 foot-pounds

   B. 9 pounds per foot

   C. 225 foot-pounds

   D. 225 pounds per foot

9. _____

10. A golf ball is hit with an initial velocity of 150 ft/sec at an angle of 29° from the horizontal. How long is the ball in the air?

10. _____

6.4–6.6

6 Chapter Quiz

DATE _____

NAME _____

Show your work on the following questions. A graphing calculator may be necessary to answer some problems. Answers should be accurate to two decimal places.

1. Point A has rectangular coordinates $(-4, 4)$. What are two sets of polar coordinates for point A, where $0 < \theta < 2\pi$?

1. _____

2. Identify the graph of the polar equation $r = a + b \cos \theta$, where $1 < \dfrac{a}{b} < 2$.

   A. cardioid
   B. rose curve
   C. limaçon with inner loop
   D. dimpled limaçon
   E. convex limaçon

2. _____

3. Write the polar equation $r = 3 \sin \theta$ in rectangular form.

3. _____

4. Plot the complex number $-2 + 3i$ in the complex plane and find its absolute value.

4. _____

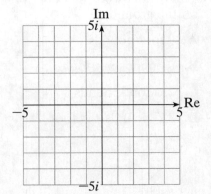

5. Write the complex number $2\sqrt{3} \left(\cos \dfrac{2\pi}{3} + i \sin \dfrac{2\pi}{3} \right)$ in standard form.

5. _____

6.4–6.6

6 Chapter Quiz (continued)   NAME

6. Find the polar equation that corresponds to the graph below.

6. _____

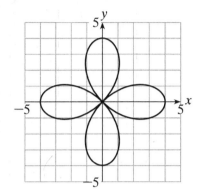

7. Draw the graph of $r = a + a \cos \theta$ where $a > 1$. (Make sure you label the scale on your graph.)

7.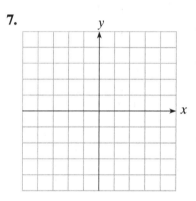

8. Draw the graph of $r^2 = 9 \sin 2\theta$.

8.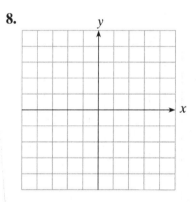

9. Find $(2 + \sqrt{5}i)^3$ using De Moivre's theorem.

   A. $-0.81 + 0.58i$
   B. $-22 + 15.65i$
   C. $0.18 + 0.98i$
   D. $0.54 + 2.95i$
   E. $15.65 - 22i$

9. _____

10. Find the fifth roots of unity.

10. _____

FORM A

6 Chapter Test

DATE _____

NAME _____

Show all work where appropriate. A graphing calculator may be necessary to answer some questions.

1. Let $A = (2, 5)$, $B = (-3, 6)$. Express the vector determined by $\overrightarrow{AB}$ as a linear combination of $\mathbf{i} = \langle 1, 0 \rangle$ and $\mathbf{j} = \langle 0, 1 \rangle$.

1. _____

2. The scalar k is equal to 4. Vectors $\mathbf{v}_1$ and $\mathbf{v}_2$ are defined as $\mathbf{v}_1 = \langle -3, 5 \rangle$ and $\mathbf{v}_2 = \langle 2, -1 \rangle$. Evaluate $\mathbf{kv}_1 + \mathbf{v}_2$.

2. _____

3. An airplane is flying on a bearing of 265° at 420 mph. A wind is blowing with a bearing of 210° at 30 mph. Find the actual speed and direction of the plane and explain how you arrived at your answer.

3. _____
_____

4. Find the unit vector in the direction of $(7, -24)$.

4. _____

5. What is the vector projection of $\mathbf{u} = \langle 7, 3 \rangle$ onto $\mathbf{v} = \langle 3, -3 \rangle$?

5. _____

6. Evaluate $3\left(\cos \dfrac{2\pi}{3} + i \sin \dfrac{2\pi}{3}\right) \cdot 4\left(\cos \dfrac{\pi}{12} + i \sin \dfrac{\pi}{12}\right)$. Express your answer in trigonometric form.

6. _____

7. Let $\mathbf{u} = \langle -1, -1 \rangle$. Find the vector $\mathbf{v}$ such that $\mathbf{u} \cdot \mathbf{v} = -6$ and $|\mathbf{v}| = \sqrt{18}$.

7. _____

FORM A
6 Chapter Test (continued)   NAME

8. Determine the rectangular coordinates of the point with polar coordinates $(8, 325°)$.

8. _____

9. Find the quotient $\dfrac{z_1}{z_2}$ for the complex numbers

$z_1 = 7(\cos 220° + i \sin 220°)$ and
$z_2 = 2(\cos 160° + i \sin 160°)$.

9. _____

10. The parametric equations $x(t) = \dfrac{1}{\sqrt{t+1}}$ and $y(t) = \dfrac{t}{t+1}$, where $t \neq -1$, represent what curve? What is the rectangular form of this curve?

10. _____

11. An NFL punter at the 15-yard line kicks a football with initial velocity of 90 feet per second at an angle of elevation of 70°. Let t be the elapsed time since the football is kicked.

(a) Write parametric equations that represent the distance the football travels. Assume that $x = 0$ represents the 15-yard line.

(b) What interval represents the possible values for t?

(c) What is the distance the ball travels in feet downfield (to the nearest whole number)?

11. (a) _____

(b) _____

(c) _____

12. Which parametric equations describe a spiral beginning at the origin for $t \geq 0$?

A. $x = t \cos t$ and $y = t \sin t$

B. $x = t \sin t$ and $y = t \sin t$

C. $x = 5 \cos t$ and $y = 5 \sin t$

D. $x = \cos t$ and $y = \sin t$

E. $x = t$ and $y = \sin t$

12. _____

FORM A

6 Chapter Test (continued)   NAME

13. The locations, given in polar coordinates, of two planes approaching the Ontario airport are (7 mi, 48°) and (5 mi, 125°). Find the distance between the planes.

13.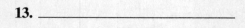

14. State the smallest θ-interval ($0 \leq \theta \leq k$) that gives a complete graph of the polar equation $r = 2 \sin 3\theta$. Sketch the graph and label the graph to show which petals are generated by positive values of r and which by negative values of r.

14.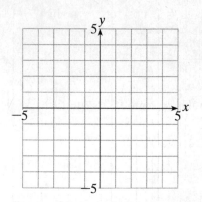

In Problems 15–16, assume that $w = 5\left(\cos \dfrac{\pi}{12} + i \sin \dfrac{\pi}{12}\right)$ is one of the six roots of a complex number z.

15. List the other five roots of z in trigonometric form.

15.

16. Determine z. Express your answer in standard form.

16. _____

58

FORM B

6 Chapter Test

DATE

NAME

Show all work where appropriate. A graphing calculator may be necessary to answer some questions.

1. Let $A = (5, 2)$, $B = (-2, 4)$. Express the vector determined by $\overrightarrow{AB}$ as a linear combination of $\mathbf{i} = \langle 1, 0 \rangle$ and $\mathbf{j} = \langle 0, 1 \rangle$.

1. _____

2. The scalar k is equal to 5. Vectors $\mathbf{v}_1$ and $\mathbf{v}_2$ are defined as $\mathbf{v}_1 = \langle 2, -6 \rangle$ and $\mathbf{v}_2 = \langle -3, 4 \rangle$. Evaluate $\mathbf{v}_1 + k\mathbf{v}_2$.

2. _____

3. An airplane is flying on a bearing of 140° at 395 mph. A wind is blowing with a bearing of 190° at 35 mph. Find the actual speed and direction of the plane and explain how you arrived at your answer.

3. _____
   _____

4. Find the unit vector in the direction of $\langle -16, 12 \rangle$.

4. _____

5. What is the vector projection of $\mathbf{u} = \langle 8, -2 \rangle$ onto $\mathbf{v} = \langle 4, -4 \rangle$?

5. _____

6. Evaluate $2\left(\cos\dfrac{\pi}{12} + i\sin\dfrac{5\pi}{12}\right) \cdot 3\left(\cos\dfrac{\pi}{4} + i\sin\dfrac{\pi}{4}\right)$.
   Express your answer in trigonometric form.

6. _____

7. Let $\mathbf{u} = \langle 1, 1 \rangle$. Find the vector $\mathbf{v}$ such that $\mathbf{u} \cdot \mathbf{v} = 8$ and $|\mathbf{v}| = \sqrt{32}$.

7. _____

FORM B

6 Chapter Test (continued)   NAME

8. Determine the rectangular coordinates of the point with polar coordinates (7, 162°).

8. _____

9. Find the quotient $\dfrac{z_1}{z_2}$ for the complex numbers

   $z_1 = 9(\cos 240° + i \sin 240°)$ and
   $z_2 = 4(\cos 105° + i \sin 105°)$.

9. _____

10. The parametric equations $x(t) = t - 2$ and $y(t) = \dfrac{t}{t - 2}$, where $t \neq 2$, represent what curve?

    What is the rectangular form of this curve?

10. _____

11. An NFL punter at the 15-yard line kicks a football with initial velocity of 95 feet per second at an angle of elevation of 65°. Let t be the elapsed time since the football is kicked.

    (a) Write parametric equations that represent the distance the football travels. Assume that $x = 0$ represents the 15-yard line.

    (b) What interval represents the possible values for t?

    (c) What is the distance the ball travels in feet downfield (to the nearest whole number)?

11. (a) _____

    (b) _____

    (c) _____

12. Which parametric equations describe a circle with its center at the origin and a radius of 7 for $0 \leq t \leq 2\pi$?

    A. $x = 7 \cos t$ and $y = 7 \sin t$

    B. $x = 7 \sin t$ and $y = 7 \sin t$

    C. $x = \cos 7t$ and $y = \sin 7t$

    D. $x = \dfrac{1}{7} \cos t$ and $y = \dfrac{1}{7} \sin t$

    E. $x = 49 \cos t$ and $y = 49 \sin t$

12. _____

FORM B

6 Chapter Test (continued)   NAME

13. The locations, given in polar coordinates, of two planes approaching the Toronto airport are (820 ft, 85°) and (460 ft, 38°). Find the distance between the planes.

13. _____

14. State the smallest θ-interval ($0 \leq \theta \leq k$) that gives a complete graph of the polar equation $r = 3 \sin 2\theta$. Sketch the graph and label the graph to show which petals are generated by positive values of r and which by negative values of r.

14.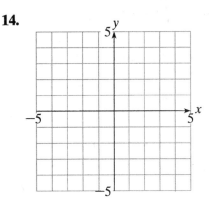

In Problems 15–16, assume that $w = 6\left(\cos \dfrac{\pi}{10} + i \sin \dfrac{\pi}{10}\right)$ is one of the five roots of a complex number z.

15. List the other four roots of z in trigonometric form.

15. _____

16. Determine z. Express your answer in standard form.

16. _____

7.1–7.2

7 Chapter Quiz

1. There are how many solutions of the system?

   $x^2 + y^2 = 25$
   $y = x^2 - 7$

   A. None **B.** One **C.** Two
   D. Three **E.** Four

   1. _____

2. Solve the system by the elimination method.

   $2x - 3y = 17$
   $4x + 5y = 1$

   2. _____

3. Solve the system by the substitution method.

   $x^2 + y^2 = 13$
   $y = x^2 - 1$

   3. _____

4. Solve the system graphically.

   $8x^2 + 32y^2 = 256$
   $x = 2y$

   4. _____

   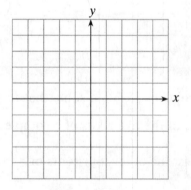

5. Find the determinant of the matrix $A = \begin{bmatrix} 1 & 5 \\ 6 & 10 \end{bmatrix}$.

   5. _____

6. A factory records its inventory in matrix X, where column 1 represents Model A of a product, column 2 represents Model B of the same product, and each row represents a color of the product. Matrix M is the price of Model A (row 1) and Model B (row 2). Compute matrix XM and give its physical interpretation.

   $X = \begin{bmatrix} 25 & 10 \\ 30 & 50 \\ 15 & 40 \\ 10 & 5 \end{bmatrix}$ and $M = \begin{bmatrix} 220 \\ 250 \end{bmatrix}$

   6. _____

7.1–7.2

7 Chapter Quiz (continued)   NAME

7. Using matrix $A = \begin{bmatrix} 3 & 7 & 4 \\ 2 & -1 & 6 \end{bmatrix}$ and

$B = \begin{bmatrix} -1 & 5 & 2 \\ 7 & 10 & 1 \\ -2 & 5 & -1 \end{bmatrix}$, which of the following is a true statement?

A. Matrix A has an inverse.

B. $A + B = \begin{bmatrix} 2 & 12 & 6 \\ 9 & 9 & 7 \\ -2 & 5 & -1 \end{bmatrix}$

C. AB exists and BA does not exist.

D. BA exists and AB does not exist.

7. _____

8. Find the determinant of the matrix

$B = \begin{bmatrix} 2 & -1 & 4 \\ 3 & 1 & -2 \\ 1/2 & 4 & 5 \end{bmatrix}$. Does matrix B have an inverse?

8. _____

9. Find the inverse of the matrix $\begin{bmatrix} 6 & -8 \\ 2 & 3 \end{bmatrix}$.

9. _____

10. Show that matrix $A = \begin{bmatrix} 1/2 & 3/2 & -1/2 \\ 0 & 1 & 0 \\ 0 & -3 & 1 \end{bmatrix}$ is the inverse

of the matrix $B = \begin{bmatrix} 2 & 0 & 1 \\ 0 & 1 & 0 \\ 0 & 3 & 1 \end{bmatrix}$.

10. _____

63

7 Chapter Quiz

7.3–7.5

1. Write the system of equations as a matrix equation $AX = B$, with A as the coefficient matrix.

   $x + 2y - 3z = 19$

   $2x - 3y + 4z = -17$

   $3x - y = 4$

   1. _____

2. Use an inverse matrix to solve the system given in Problem 1.

   2. _____

3. Find the solution to the system of equations represented by the augmented matrix.

   $$\begin{bmatrix} 1 & 3 & -4 & -13 \\ 0 & 1 & 2 & 4 \\ 0 & 0 & 1 & 3 \end{bmatrix}$$

   3. _____

4. The system below has an infinite number of solutions. Find the solution set.

   $x + y - 5z = 3$

   $x - 2z = 1$

   $2x - y - z = 1$

   4. _____

5. Write the partial fraction decomposition of $\dfrac{3x^2 + 4}{(x^2 + 1)^2}$.

   5. _____

6. When writing the solution to the partial fraction decomposition of $\dfrac{x^3}{x^2 - x - 2}$, what is the first step of this solution?

   6. _____

   A. $\dfrac{x^3}{x^2 - x - 2} = \dfrac{A}{x - 2} + \dfrac{B}{x + 1}$

   B. $\dfrac{x^3}{x^2 - x - 2} = \dfrac{Ax + B}{x - 2} + \dfrac{C}{x + 1}$

   C. $\dfrac{x^3}{x^2 - x - 2} = x + 1 + \dfrac{3x + 2}{x^2 - x - 2}$

   D. $\dfrac{x^3}{x^2 - x - 2} = x + \dfrac{3x + 2}{x^2 - x - 2}$

   E. $\dfrac{x^3}{x^2 - x - 2} = 1 + \dfrac{3x + 2}{x^2 - x - 2}$

7.3–7.5

7 Chapter Quiz (continued)   NAME

7. Determine the inequality corresponding to the graph below.

7. _____

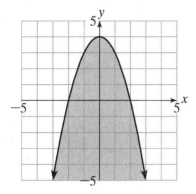

8. Graph the inequality $x^2 + y^2 \geq 4$.

8.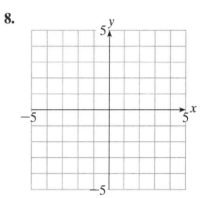

9. Draw a graph representing the solution to the system of inequalities.

$16x^2 + 25y^2 \leq 400$

$y \geq (1/3)x - 1$

9.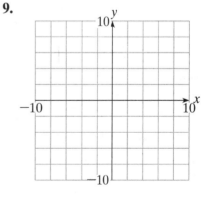

10. Solve the linear programming problem: Maximize $C = 2x + 5y$ subject to

$x + 2y \leq 10$
$3x + 2y \leq 18$
$x \geq 0$
$y \geq 0$

A. 12   B. 23   C. 25   D. 32

10. _____

65

FORM A   DATE

7 Chapter Test   NAME

Directions:

Show all work where appropriate. A graphing calculator may be necessary to answer some questions.

1. Solve the system by substitution.   1. _____

   $x - y = 3$

   $y = x^2 - 5$

2. Solve the system by elimination.   2. _____

   $x - y = -2$

   $x + 2y = 7$

3. Solve the system graphically. Round answers to the nearest hundredth.   3. _____

   $y = -1 + 3x^2 + 2x^3$

   $y = 1 + 4x$

4. Tell what elementary row operation can be applied to   4. _____

   $\begin{bmatrix} 3 & 6 & -2 & 5 \\ 4 & 7 & 5 & -2 \\ -3 & 2 & 0 & 3 \end{bmatrix}$ to obtain $\begin{bmatrix} 15 & 27 & 13 & -1 \\ 4 & 7 & 5 & -2 \\ -3 & 2 & 0 & 3 \end{bmatrix}$.

   A. $5R_1$

   B. $3R_2 + R_1$

   C. $3R_1 + R_2$

   D. $3R_2 - R_3$

5. Use Gaussian elimination to solve the system.   5. _____

   $x + 2y - 3z = -7$

   $2x - 3y + z = 14$

   $4x + y - 2z = 3$

6. Use Gaussian elimination to solve the system.   6. _____

   $2x + 3y - 12z = 1$

   $x - 2y + z = 4$

   $4x + y - 14z = 7$

FORM A

7 Chapter Test (continued)   NAME

7. Determine the number of solutions to the augmented matrix.

$$\begin{bmatrix} 1 & 2 & 0 & 1 & 4 \\ 0 & 1 & -1 & -3 & 2 \\ 0 & 0 & 1 & 3 & 5 \\ 0 & 0 & 0 & 0 & 1 \end{bmatrix}$$

A. infinitely many

B. 2 solutions

C. 1 solution

D. no solutions

7. _____

8. A company's inventory for two models of computers and two models of palm-pilots is given by the matrix $\begin{bmatrix} 260 & 400 \\ 300 & 180 \end{bmatrix}$, where row 1 represents the computers, row 2 represents the palm-pilots, column 1 represents Model A, and column 2 represents Model B. If the company wants to increase inventory by 15% in the next month, what will be the projected matrix for the inventory? Use matrices and explain.

8. $\begin{bmatrix} & \end{bmatrix}$ _____

_____

9. Write the system of equations for the augmented matrix. Do not solve.

$$\begin{bmatrix} 2 & 1 & 0 & 3 \\ -1 & 3 & 4 & 0 \\ 0 & -2 & 1 & 5 \end{bmatrix}$$

9. _____

_____

_____

10. The inverse of matrix $\begin{bmatrix} 1 & 1 & 0 \\ 1 & 2 & 1 \\ 2 & 3 & 2 \end{bmatrix}$ is $\begin{bmatrix} 1 & -2 & 1 \\ 0 & 2 & -1 \\ -1 & -1 & 1 \end{bmatrix}$.

10. _____

Solve the following system by the method of inverse matrix.

$$x + y = 15$$
$$x + 2y + z = 20$$
$$2x + 3y + 2z = 12$$

FORM A
7 Chapter Test (continued)   NAME

11. Jose invested a total of $40,000 in three different investments that yield 6%, 8%, and 10% per year, respectively. The total income per year from the three investments is $3,240. If the income from the 8% investment is half the income of the 10% investment, what amount did Jose invest in each investment?

    (a) Write the system of equations that will solve this problem. Do not solve.

    (b) Write the system as a matrix equation $AX = B$. Do not solve.

11. (a)

    (b)

12. Find A^{-1} of the matrix $A = \begin{bmatrix} 1 & 0 & 0 \\ -2 & 1 & 5 \\ 4 & 0 & 1 \end{bmatrix}$.

12. _____

13. What is the partial fraction decomposition of $\dfrac{5x}{x^2 + x - 6}$?

13. _____

14. Solve the system of inequalities graphically.
$$x^2 + y^2 = 36$$
$$100x^2 + 18y^2 \geq 1600$$

14.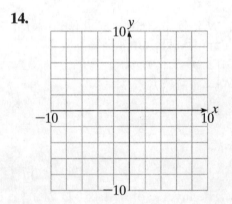

15. Solve the linear programming problem.
Maximize $C = -3x + 5y$ subject to:
$$x - y \geq -3$$
$$2x + y \leq 12$$
$$x \geq 0$$
$$y \geq 0$$

15. $C =$ _____

    at (x, y) _____

FORM B

7 Chapter Test

DATE _____

NAME _____

Directions:
Show all work where appropriate. A graphing calculator may be necessary to answer some questions.

1. Solve the system by substitution.

   $x - y = 4$

   $y = x^2 - 6$

   1. _____

2. Solve the system by elimination.

   $x - 2y = 4$

   $x + y = 1$

   2. _____

3. Solve the system graphically. Round answers to the nearest hundredth.

   $y = 4 - 7x^2 + 3x^3$

   $y = -1 + 2x$

   3. _____

4. Tell what elementary row operation can be applied to
   $\begin{bmatrix} 3 & 6 & -2 & 5 \\ 4 & 7 & 5 & -2 \\ -3 & 2 & 0 & 3 \end{bmatrix}$ to obtain $\begin{bmatrix} 3 & 6 & -2 & 5 \\ 10 & 3 & 5 & -8 \\ -3 & 2 & 0 & 3 \end{bmatrix}$.

   A. $-2R_2 + R_3$

   B. $\dfrac{10}{4} R_2$

   C. $-2R_3 + R_2$

   D. $2R_1 + R_2$

   4. _____

5. Use Gaussian elimination to solve the system.

   $5x - 2y - 4z = 24$

   $4x - y - 5z = 18$

   $x + 3y - 11z = -2$

   5. _____

6. Use Gaussian elimination to solve the system.

   $x + 2y - 3z = 3$

   $4x + y + 2z = -2$

   $2x - 3y + 8z = 5$

   6. _____

FORM B

7 Chapter Test (continued)   NAME

7. Determine the number of solutions to the augmented matrix.

$$\begin{bmatrix} 1 & 0 & 0 & 4 & 3 \\ 0 & 1 & 0 & 3 & -1 \\ 0 & 0 & 1 & 2 & 1 \\ 0 & 0 & 0 & 1 & 0 \end{bmatrix}$$

A. infinitely many

B. 2 solutions

C. 1 solution

D. no solutions

7. _____

8. A company's inventory for two models of televisions and two models of radios is given by the matrix $\begin{bmatrix} 360 & 300 \\ 200 & 160 \end{bmatrix}$, where row 1 represents the televisions, row 2 represents the radios, column 1 represents Model A, and column 2 represents Model B. If the company wants to increase inventory by 20% in the next month, what will be the projected matrix for the inventory? Use matrices and explain.

8. $\begin{bmatrix} \end{bmatrix}$ _____

9. Write the system of equations for the augmented matrix. Do not solve.

$$\begin{bmatrix} 0 & 4 & 3 & -1 \\ 5 & 1 & -2 & 0 \\ 3 & 0 & 1 & 2 \end{bmatrix}$$

9. _____

10. The inverse of matrix $\begin{bmatrix} 2 & 1 & -1 \\ 2 & 2 & -1 \\ -1 & -1 & 1 \end{bmatrix}$ is $\begin{bmatrix} 1 & 0 & 1 \\ -1 & 1 & 0 \\ 0 & 1 & 2 \end{bmatrix}$.

Solve the following system by the method of inverse matrix.

$2x + y - z = 2$

$2x + 2y - z = 4$

$-x - y + z = -1$

10. _____

FORM B

7 Chapter Test (continued) NAME

11. Connor invested a total of $50,000 in three different investments that yield 5%, 7%, and 10% per year, respectively. The total income per year from the three investments is $3,764. If the income from the 7% investment is half the income of the 10% investment, what amount did Jose invest in each investment?

(a) Write the system of equations that will solve this problem. Do not solve.

(b) Write the system as a matrix equation $AX = B$. Do not solve.

11. (a) _____

_____

_____

(b) _____

12. Find A^{-1} of the matrix $A = \begin{bmatrix} 1 & 0 & 0 \\ -3 & 1 & 6 \\ 5 & 0 & 1 \end{bmatrix}$.

12. _____

13. What is the partial fraction decomposition of $\dfrac{2x}{x^2 - 2x - 3}$?

13. _____

14. Solve the system of inequalities graphically.

$x^2 + y^2 \leq 25$

$xy \leq 2$

14.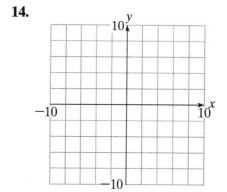

15. Solve the linear programming problem.
Maximize $C = 4x - 3y$ subject to:

$x + 4y \leq 32$

$2x - y \leq 10$

$x \geq 0$

$y \geq 0$

15. $C = $ _____

at (x, y) _____

8.1–8.3

8 Chapter Quiz

1. What is the vertex of the parabola $5x^2 - 40x - y + 78 = 0$?

2. What is the equation, in standard form, of the parabola with focus $(3, 0)$ and directrix $x = -3$?

3. What is the equation of a parabola with vertex $(-4, -4)$ that passes through $(0, 0)$ and has its axis of symmetry parallel to the y-axis?

4. Sketch a graph of the equation $\dfrac{x^2}{9} + \dfrac{y^2}{16} = 1$.

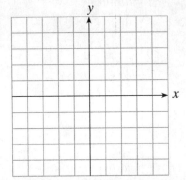

5. Find the asymptotes of the hyperbola $\dfrac{x^2}{81} - \dfrac{y^2}{25} = 1$.

   A. $y = \dfrac{5}{9}x$ and $y = -\dfrac{5}{9}x$

   B. $y = \dfrac{5}{9}x + 10.30$ and $y = -\dfrac{5}{9}x + 10.30$

   C. $y = \dfrac{9}{5}x$ and $y = -\dfrac{9}{5}x$

   D. $y = 2.06x$ and $y = -2.06x$

6. Find an equation in standard form for the ellipse whose major axis endpoints are $(-5, 2)$ and $(3, 2)$ and whose minor axis has length 6.

7. Explain how you can use the eccentricity to determine whether the graph of a given quadratic equation is an ellipse or a circle.

8. Find the foci of the ellipse $\dfrac{(x-2)^2}{64} + \dfrac{(y+3)^2}{25} = 1$.

9. Which of the following are the vertices of the hyperbola $\dfrac{(x-3)^2}{25} - \dfrac{(y+5)^2}{169} = 1$?

   A. $(3, 8)$ and $(3, -18)$ B. $(8, -5)$ and $(-2, -5)$

   C. $(3, 8)$ and $(3, -13)$ D. $(16, -5)$ and $(-10, -5)$

10. Find an equation in standard form for the hyperbola centered at $(1, -4)$, with one focus at $(7, -4)$ and eccentricity $e = 2$.

8.4–8.6

8 Chapter Quiz

DATE

NAME

1. What type of conic is the graph of $r = 3/(1 - \cos \theta)$?

1. _____

2. Draw the graph of $r = 4/(2 + \cos \theta)$.

2.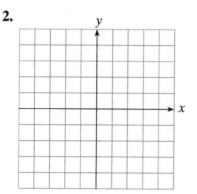

3. Find a polar equation for the conic with focus at $(0, 0)$, eccentricity $4/3$, and directrix $y = 7$.

3. _____

4. Determine the directrix of the conic with polar form $r = 6/(3 - \cos \theta)$.

4. _____

5. Use the discriminant $B^2 - 4AC$ to determine if $3x^2 - 12xy + 4y^2 + x - 5y - 4 = 0$ is an ellipse, a parabola, or a hyperbola.

5. _____

6. After translation of coordinate axes, a quadratic equation becomes $\dfrac{(x')^2}{16} + \dfrac{(y')^2}{9} = 1$. If the translation had the effect of moving the point $(x, y) = (1, -2)$ to the point $(x', y') = (0, 0)$, what was the original equation in xy coordinates?

6. _____

A. $\dfrac{x^2}{16} + \dfrac{y^2}{9} = 1$

B. $\dfrac{(x + 1)^2}{16} + \dfrac{(y - 2)^2}{9} = 1$

C. $\dfrac{(x - 1)^2}{16} + \dfrac{(y + 2)^2}{9} = 1$

D. $\dfrac{(x' - 1)^2}{16} + \dfrac{(y' + 2)^2}{9} = 1$

7. Find the coordinates of the point $P(x, y) = (3, 4)$ in the coordinate system $x'y'$ obtained by a rotation of $\alpha = \pi/3$.

7. _____

8 Chapter Quiz (continued)   NAME

8. Which of the following is the graph of the plane $2x + 6y + z = 1$?

8. _____

A.
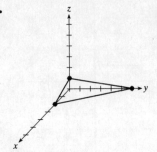

B.
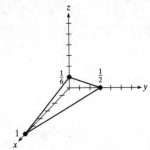

C.
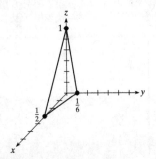

D.
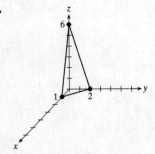

9. For the points $A\,(-2, 4, 6)$, $B\,(0, 6, -2)$ and $C\,(4, -8, 6)$, what is the distance from point A to the midpoint of line segment BC?

9. _____

10. Write the parametric equations for the line through the points $(-1, 2, 4)$ and $(6, 0, -3)$.

10. _____

FORM A

8 Chapter Test

DATE

NAME

1. The point $(-3, -2, 10)$ is what distance from the xz-plane?

   A. -3
   B. -2
   C. 2
   D. 3
   E. 10

   1. _____

2. What is the equation of the parabola with vertex $(-4, -4)$ that passes through $(0, 0)$ and has its axis of symmetry parallel to the x-axis?

   2. _____

3. For the hyperbola $\dfrac{(y-5)^2}{9} - \dfrac{(x-3)^2}{25} = 1$, find the center, foci, endpoints of the transverse axis, and the equations of the asymptotes.

   3. Center: _____

   Foci: _____

   Endpoints: _____

   Asymptotes: _____

4. Sketch the graph of the hyperbola in Problem 3.

   4.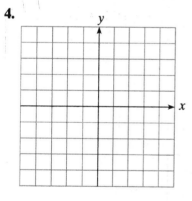

5. Find the vertex of the parabola $y = 2x^2 - 12x + 23$.

   5. _____

6. Write the equation of the conic with focus $(3, 4)$ and directrix $x = 1$.

   6. _____

7. Find a polar equation for an ellipse with a focus at $(0, 0)$ if the endpoints of its major axis have polar coordinates $(5, \pi/2)$ and $(3, 3\pi/2)$.

   7. _____

FORM A

8 Chapter Test (continued)   NAME

8. In a whispering gallery, any sound at one focus of an ellipse reflects off the ellipse directly to the other focus. The figure shows an elliptical whispering gallery, where x is the center of the ellipse. What is the distance that sound travels from focus F_1 to focus F_2?

8. _____

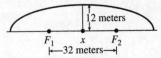

9. Identify the conic $r = \dfrac{24}{5 + 8\cos\theta}$ without drawing its graph. State the eccentricity and directrix.

9. _____

10. A telecommunications satellite is a paraboloid of revolution with a diameter of 8 ft and depth of 3 ft. How far from the vertex should the receiving antenna be placed?

10. _____

11. Identify the type of conic given by the equation $2x^2 - xy - 3y^2 - 2x + 4y - 6 = 0$.

11. _____

12. For the equation in Problem 11, solve for y to obtain equation(s) that could be entered into your grapher.

12. _____

13. Find the equation for the conic in standard form. Identify the conic.
$x = 5 + 2\cos t, y = -2 + \sin t, 0 \le t \le 2\pi$.

13. _____

14. What is the magnitude of the vector $\mathbf{v} = \langle 1, -4, 2 \rangle$?

14. _____

15. In three-dimensional space, what are the parametric equations for the line through the points $(2, -1, 3)$ and $(-1, 4, 0)$?

15. _____

16. In three-dimensional space, what is the equation of the set of points that are 12 units from the point $(1, 5, -4)$?

16. _____

FORM B
8 Chapter Test

1. The point $(-5, -4, 8)$ is what distance from the yz-plane?

   A. -5
   B. -4
   C. 4
   D. 5
   E. 8

   1. _____

2. What is the equation of the parabola with vertex $(-2, -2)$ that passes through $(0, 0)$ and has its axis of symmetry parallel to the x-axis?

   2. _____

3. For the hyperbola $\dfrac{(x+2)^2}{4} - \dfrac{(y-5)^2}{16} = 1$, find the center, foci, endpoints of the transverse axis, and the equations of the asymptotes.

   3. Center: _____
   Foci: _____
   Endpoints: _____
   Asymptotes: _____

4. Sketch the graph of the hyperbola in Problem 3.

   4.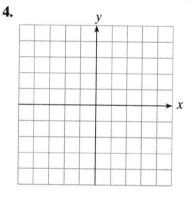

5. Find the vertex of the parabola $y = 4x^2 + 8x - 6$.

   5. _____

6. Write the equation of the conic with focus $(2, -3)$ and directrix $y = 5$.

   6. _____

7. Find a polar equation for an ellipse with a focus at $(0, 0)$ if the endpoints of its major axis have polar coordinates $(5, \pi)$ and $(3, 2\pi)$.

   7. _____

77

FORM B

8 Chapter Test (continued) NAME

8. In a whispering gallery, any sound at one focus of an ellipse reflects off the ellipse directly to the other focus. The figure shows an elliptical whispering gallery, where x is the center of the ellipse. What is the distance that sound travels from focus F_1 to focus F_2?

8. _____

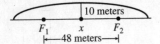

9. Identify the conic without drawing its graph. State the eccentricity and directrix.

$$r = \frac{20}{7 - 4\sin\theta}$$

9. _____

10. A telecommunications satellite is a paraboloid of revolution with a diameter of 6 ft and depth of 3 ft. How far from the vertex should the receiving antenna be placed?

10. _____

11. Identify the type of conic given by the equation $2x^2 - xy + 3y^2 - x - 4y - 6 = 0$.

11. _____

12. For the equation in Problem 11, solve for y to obtain equation(s) that could be entered into your grapher.

12. _____

13. Find the equation for the conic in standard form. Identify the conic.
$x = 2 + 2\cos t, y = -2 + 2\sin t, 0 \leq t \leq 2\pi$.

13. _____

14. What is the magnitude of the vector $\mathbf{v} = \langle 2, -1, 3 \rangle$?

14. _____

15. In three-dimensional space, what are the parametric equations for the line through the points $(4, -2, 3)$ and $(-1, 1, 0)$?

15. _____

16. In three-dimensional space, what is the equation of the set of points that are 8 units from the point $(-3, 5, 2)$?

16. _____

9.1–9.3

9 Chapter Quiz

DATE

NAME

1. A computer password is 6 entries long. If the password must have 3 letters followed by 3 numbers with repetition of numbers and/or letters permitted, how many different passwords are possible?

1. _____

2. There are 85 students in a college mathematics class. Five students are chosen to work on a project together. How many different groups of five students are there in this class?

2. _____

3. A computer randomly selects a number from 1 through 50. Two selections are made by the computer to create an ordered pair of numbers. How many elements are in the sample space for this experiment?

   A. 2,500   B. 2,450   C. 2,000
   D. 100   E. 50

3. _____

4. For the experiment in Problem 3, what is the probability that the computer will not select the same number in both selections?

4. _____

5. A fair coin is tossed five times. Find the probability of tossing exactly two tails in those five tosses.

5. _____

9.1–9.3

9 Chapter Quiz (continued)   NAME

6. Suppose there is a 40% chance of rain tomorrow. If it rains, there is a 20% chance that all of the rides at an amusement park will be operating. If it doesn't rain, there is a 90% chance all of the rides will be operating. What is the probability that all of the rides will be operating tomorrow?

6. _____

7. Assume that the probability that a newborn child is a female is 50%. In a family of four children, what is the probability that

  (a) all the children are girls?

  (b) at least two of the children are boys?

7. (a) _____

   (b) _____

8. Expand $(5x - 8y)^5$.

8. _____
_____

9. Find the fifteenth term of $(x + 2y)^{20}$, where x^{20} is labelled the zeroth term.

9. _____

10. Find the coefficient of the x^3y^4 term in the expansion of $(x + y)^7$.

  A. 3   B. 4   C. 7!

  D. 35   E. 210

10. _____

9.4–9.6

9 Chapter Quiz

1. The third and fifth terms of an arithmetic sequence are 2 and 32, respectively. Find explicit and recursive formulas for the sequence.

   1. Explicit: _____
      Recursive: _____

2. The third and fifth terms of a geometric sequence are 2 and 32, respectively. Find explicit and recursive formulas for the sequence.

   2. Explicit: _____
      Recursive: _____

3. The sum of the squares of the first n positive odd integers is $\dfrac{(2n-1)(2n)(2n+1)}{6}$. Use this expression to find the sum $1^2 + 3^2 + 5^2 + \cdots + 75^2$.

   A. 5628   B. 73,150   C. 1425
   D. 7,700,625   E. 5776

   3. _____

4. Find the sum of the first 12 terms of the arithmetic sequence:

   28, 22, 16, 10, ...

   4. _____

5. Do the following sequences converge or diverge? If they converge, state the limit.

   (a) $\left\{\dfrac{5n}{2n+1}\right\}$

   (b) $\left\{\dfrac{n^3+1}{2n+1}\right\}$

   (c) $\left\{\dfrac{3n^3}{4n^5+1}\right\}$

   5. (a) _____
      (b) _____
      (c) _____

9.4–9.6

9 Chapter Quiz (continued)   NAME

6. Find the sum of the infinite series $\sum_{k=1}^{\infty} 4\left(\frac{1}{5}\right)^{k-1}$.

6. _____

7. Write the statements P_1, P_k, and P_{k+1} for the equation $1(1!) + 2(2!) + \cdots + n(n!) = (n+1)! - 1$ as if you were writing a proof by mathematical induction. Do not write a proof.

7. $P_1 =$ _____

$P_k =$ _____

$P_{k+1} =$ _____

8. Sally deposits $100 at the end of each month into an account that pays 6% annual interest compounded monthly. After 5 years of deposits, what is the account balance?

8. _____

9. Determine the sixth partial sum of the geometric sequence $5, \frac{5}{4}, \frac{5}{16}, \ldots$

9. _____

  A. $\dfrac{5}{4096}$ **B.** $\dfrac{1}{1024}$ **C.** $\dfrac{5}{1024}$

  D. 6.48   **E.** 6.67

10. Prove the following statement by mathematical induction:

$3 + 7 + 11 + \cdots + (4n - 1) = n(2n + 1)$

10. _____

_____

_____

_____

82

9.7–9.8

9 Chapter Quiz

DATE _____

NAME _____

1. Twenty-four students participated in a basketball contest. Each student took five shots. The number of baskets scored by each student is shown below. Find the mean, median, and mode for the data.

4	3	3	4	5	3
2	3	2	5	5	5
1	2	5	3	3	4
0	2	5	1	3	5

1. Mean = _____
   Median = _____
   Mode = _____

2. Find the standard deviation and variance of the scores on a final exam.

   65, 88, 83, 80, 78, 90, 92, 85, 74, 70, 95, 62

2. Standard dev. = _____
   Variance = _____

3. Make a histogram of the data, using interval widths of 10 points.

Student	Al	Barry	Chen	Doris	Ella	Fred	Geoff
Score	69	85	73	78	62	82	55

Student	Harold	Ingrid	Jay	Kay	Leon	Matt	Nguyen
Score	79	67	84	75	95	87	91

3.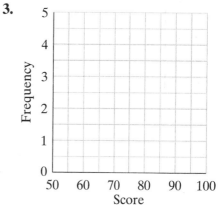

4. Find the five-number summary and the range for the data in the stemplot below.

Stem	Leaf
3	5 5 6
4	2 3 3 6 8
5	2 3 5 5 6 7
6	3 4 4 4 5 5 6
7	2 3 4 5 5 6 7 7

4. _____
   Range = _____

5. Sketch a boxplot for the data in Problem 4.

5. ←—+—+—+—+—+—→ x
     30 40 50 60 70 80

83

9 Chapter Quiz (continued)

6. The table below shows the tuition fee for one semester at a certain university.

Year	2002	2003	2004	2005	2006
Cost	5500	5600	6000	6300	6700

Complete a line graph that shows this university's tuition fee trend.

6.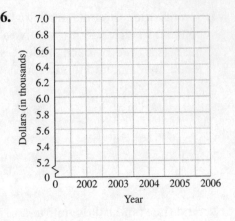

7. If the scores on a certain exam are normally distributed with a mean of 79 and a standard deviation of 7, then 95% of the scores will fall in which interval?

   A. [72, 86]   **B.** [58, 100]   **C.** [68.5, 89.5]
   D. [65, 93]   **E.** [75, 83]

7. _____

8. A machine fills 16-ounce soda cans with a mean of 16.06 ounces of soda and a standard deviation of 0.03. Approximately what percentage of the cans will actually contain less than the advertised less than the advertised 16 ounces of soda?

   A. 2.5%   **B.** 5%   **C.** 10%
   D. 12.5%   **E.** 15%

8. _____

9. A student's grade in a particular class is determined as follows:

   mean of three class exams—45%
   final exam—25%
   project—30%

   If a student scored 85, 88, and 76 on the three exams, 90 on the final exam, and 86 on the project, what is the student's grade in the class? (Round your answer to the nearest whole number.)

9. _____

10. A student has an average of 72% on 7 tests. What average score will this student need to achieve on the next 3 tests in order to raise her overall average to 75%?

10. _____

FORM A

9 Chapter Test

DATE _____

NAME _____

Directions: Show all work where appropriate. A graphing calculator may be necessary to answer some questions.

1. The sequence 2, 6, 18, 54, ... is geometric.

   (a) What is the common ratio of the sequence?

   (b) What is a recursive rule for the nth term ($n \geq 2$) of the sequence?

   1. (a) _____

   (b) _____

2. Which one of the following sequences is divergent?

   A. $\left\{\dfrac{n + 200}{n}\right\}$ B. $\{e^{-n}\}$ C. $\{\sqrt[3]{n}\}$

   D. $\{n^{-5}\}$ E. $\left\{\dfrac{100n + 1}{n}\right\}$

   2. _____

3. Write the series in summation notation and find the sum, assuming the suggested pattern continues.

   $6 + 2 + \dfrac{2}{3} + \dfrac{2}{9} + \ldots + \dfrac{2}{729}$

   3. _____

4. For each of the sequences below, determine whether the infinite geometric series converges. If it does converge, what is the sum?

   (a) $\dfrac{5}{4} + \dfrac{5}{16} + \dfrac{5}{64} + \ldots$

   (b) $\dfrac{1}{96} + \dfrac{1}{32} + \dfrac{3}{32} + \dfrac{9}{32} + \ldots$

   4. (a) _____

   (b) _____

5. Using the binomial theorem, expand the binomial $(2x + y)^5$.

   5. _____

6. Use the formulas $\sum_{k=1}^{n} k = \dfrac{n(n + 1)}{2}$ and $\sum_{k=1}^{n} k^2 = \dfrac{n(n + 1)(2n + 1)}{6}$ to find the sum in terms of n. $\sum_{k=1}^{n} (3k^2 - 5k + 7)$

   6. _____

7. How many different groups of officers—president, vice-president, treasurer, and secretary—can be formed from an organization with 35 members?

   7. _____

8. A hoagie consists of a roll, one type of meat, one type of cheese, and three different toppings. If there are 5 kinds of rolls, 5 kinds of meat, 4 kinds of cheese, and 9 toppings from which to choose, how many different hoagies can be built?

   8. _____

85

FORM A

9 Chapter Test (continued)   NAME

9. What is the probability of obtaining exactly four heads in eleven tosses of a fair coin?

    A. $330\left(\frac{1}{2}\right)^{11}$ B. $\left(\frac{1}{2}\right)^{4}$ C. $7920\left(\frac{1}{2}\right)^{11}$

    D. $4!\left(\frac{1}{2}\right)^{4}$ E. $11!\left(\frac{1}{2}\right)^{11}$

    9. _____

10. Two dice are rolled. Find the probability of rolling a sum greater than 6.

    10. _____

11. Use the definition of combinations $\binom{n}{r}$ to show that $\binom{6}{r-1} + \binom{6}{r} = \binom{7}{r}$, where $1 \leq r \leq 6$.

    11. _____

Questions 12–15 refer to the following data set. {51, 57, 43, 65, 72, 39, 56, 61, 48, 49, 37, 44, 68, 75, 52, 56, 41, 63}

12. Construct a stemplot for the data.

    12.

13. In the space below, construct a frequency table for the data. Then draw the histogram. Use intervals of 10.

    13.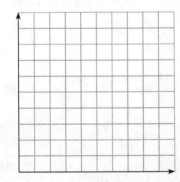

14. Find the mean, median, and variance. If necessary, round to the nearest hundredth.

    14. Mean: _____
    Median: _____
    Variance: _____

15. Construct a boxplot for the data.

    15. ←——+—+—+—+—+—→ x

16. Prove the following statement by mathematical induction.

    $3 + 7 + 11 + \ldots + (4n - 1) = n(2n + 1)$

    16. _____

86

FORM B   DATE

9 Chapter Test   NAME

Directions: Show all work where appropriate. A graphing calculator may be necessary to answer some questions.

1. The sequence 25, 21, 17, 13, ... is arithmetic.

    (a) What is the common difference of the sequence?

    (b) What is a recursive rule for the nth term ($n \geq 2$) of the sequence?

    1. (a) _____
       (b) _____

2. Which one of the following sequences is divergent?

    A. $\left\{\dfrac{3n + 200}{n}\right\}$ **B.** $\{e^{-n^2-1}\}$ **C.** $\{n^{-3}\}$

    D. $\{\sqrt{n}\}$ **E.** $\left\{\dfrac{50n + 200}{n}\right\}$

    2. _____

3. Write the series in summation notation and find the sum, assuming the suggested pattern continues.

    $-5 + 2 + 9 + 16 + \ldots + 177$

    3. _____

4. For each of the sequences below, determine whether the infinite geometric series converges. If it does converge, what is the sum?

    (a) $\dfrac{5}{3} + \dfrac{5}{9} + \dfrac{5}{27} + \ldots$

    (b) $\dfrac{1}{96} + \dfrac{1}{16} + \dfrac{3}{8} + \ldots$

    4. (a) _____
       (b) _____

5. Using the binomial theorem, expand the binomial $(3x + y)^4$.

    5. _____

6. Use the formulas $\displaystyle\sum_{k=1}^{n} k = \dfrac{n(n+1)}{2}$ and $\displaystyle\sum_{k=1}^{n} k^2 = \dfrac{n(n+1)(2n+1)}{6}$ to find the sum in terms of n. $\displaystyle\sum_{k=1}^{n}(4k^2 + 2k - 9)$

    6. _____

7. How many different groups of officers—president, vice-president, treasurer, and secretary—can be formed from an organization with 45 members?

    7. _____

8. A hoagie consists of a roll, two types of meat, one type of cheese, and four different toppings. If there are 6 kinds of rolls, 7 kinds of meat, 4 kinds of cheese, and 10 toppings from which to choose, how many different hoagies can be built?

    8. _____

87

FORM B

9 Chapter Test (continued)   NAME

9. Find the probability of obtaining exactly six heads in ten tosses of a fair coin.

   A. $10!\left(\dfrac{1}{2}\right)^{10}$ **B.** $\left(\dfrac{1}{2}\right)^{6}$ **C.** $151{,}200\left(\dfrac{1}{2}\right)^{10}$

   D. $6!\left(\dfrac{1}{2}\right)^{6}$ **E.** $210\left(\dfrac{1}{2}\right)^{10}$

9. _____

10. Two dice are rolled. Find the probability of rolling a sum greater than 8.

10. _____

11. Use the definition of combinations $\binom{n}{r}$ to show that $\binom{k}{7} + \binom{k}{8} = \binom{k+1}{8}$, where $k \geq 8$.

11. _____
_____
_____

Questions 12–15 refer to the following data set. {27, 42, 33, 45, 52, 18, 37, 37, 29, 42, 58, 27, 19, 56, 45, 38}

12. Construct a stemplot for the data.

12. _____

13. In the space below, construct a frequency table for the data. Then draw the histogram. Use intervals of 10.

13.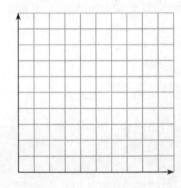

14. Find the mean, median, and variance. If necessary, round to the nearest hundredth.

14. Mean: _____
    Median: _____
    Variance: _____

15. Construct a boxplot for the data.

15. ←—+—+—+—+—+—+—→ x

16. Prove the following statement by mathematical induction.
    $1 + 7 + 13 + \ldots + (6n - 5) = n(3n - 2)$

16. _____
_____
_____

10.1–10.4

10 Chapter Quiz

1. What is the average rate of change of $f(x) = \sqrt{3x - 1}$ over the interval $[1.9, 2.1]$?

   1. _____

2. For the function $f(x) = 4 + 3x - 2x^2$:

   (a) Use the definition of the derivative at a point to find the slope of the tangent line of the function at $x = 1$.

   (b) What is the equation of the tangent line in part a?

   2. (a) _____
      (b) _____

3. What is the instantaneous rate of change of $f(x) = 2x^2$ at $x = 1$?

   A. 2 B. 0
   C. 4 D. 1

   3. _____

4. What is the derivative of $f(x) = 4x^2$?

   4. _____

5. Partition the interval $[1, 5]$ into eight equal subintervals. List the eight subintervals.

   5. _____
      _____

6. Draw the graph of $f(x) = \dfrac{1}{2}x$ over the interval $[0, 4]$. On the graph, show and shade the rectangles that would be used to approximate the area under the curve by the right rectangle approximation method using 4 subintervals.

   6.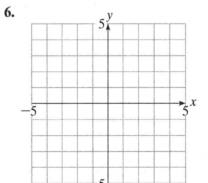

7. Compute the integral $\displaystyle\int_{2}^{5} 4x\, dx$ by computing a geometric area. (Hint: Graph the function first.)

   7. _____

FORM A

10 Chapter Test

DATE

NAME

1. Find the average rate of change of $f(x) = 3 \cos x$ over the interval $\left[\dfrac{\pi}{3}, \dfrac{2\pi}{3}\right]$.

   A. -3
   B. -9π
   C. $-\dfrac{9}{\pi}$
   D. $\dfrac{\pi}{3}$

1. _____

2. Find the instantaneous rate of change of $f(x) = 5x^2$ at $x = 3$.

2. _____

3. The following table lists the population statistics for a certain city.

Year	Population
1975	695,854
1985	720,928
1995	742,547
2005	750,370

   What is the average rate of change in the population with respect to time between 1975 and 2005?

3. _____

4. Find the derivative of $f(x) = 7x + 10$.

4. _____

5. For the function $f(x) = \dfrac{3}{x - 4}$:

   (a) Find the slope of the function at $x = 1$ using the definition of the derivative.

   (b) What is the equation of the tangent line to the curve $f(x)$ at $x = 1$?

5. (a) _____
   (b) _____

6. A coin is dropped from a window. Find the coin's
   (a) average speed during the first 4 sec of fall and
   (b) instantaneous speed at $t = 4$.

6. (a) _____
   (b) _____

90

FORM A

10 Chapter Test (continued)   NAME

7. Determine the following limits, if they exist.

(a) $\lim_{x \to -2} x(x-2)^2$

(b) $\lim_{x \to 3} \dfrac{x^2 - 7x + 12}{x^2 - 4x + 3}$

(c) $\lim_{x \to \infty} 5x - 2$

(d) $\lim_{x \to 0} \dfrac{\sin 5x}{x^2 - x}$

(e) $\lim_{x \to 3} \dfrac{x-3}{x^2 + 5}$

(f) $\lim_{x \to 0} \dfrac{2}{x}$

7. (a) _____
(b) _____
(c) _____
(d) _____
(e) _____
(f) _____

8. $\lim_{x \to -4} \dfrac{x^3 + 64}{x + 4}$

(a) Explain why direct substitution cannot be used to find the limit.

(b) Find the limit algebraically, if it exists.

8. (a) _____
(b) _____

9. Use the graph below to find the limits or explain why the limits do not exist.

(a) $\lim_{x \to 2^-} f(x)$

(b) $\lim_{x \to 2^+} f(x)$

(c) $\lim_{x \to 2} f(x)$

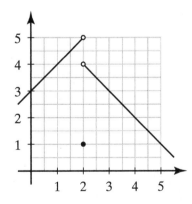

9. (a) _____
(b) _____
(c) _____

10. At what points c in the domain of $f(x)$ does $\lim_{x \to c} f(x)$ exist?

$$f(x) = \begin{cases} -x^2 & -2 \le x < 1 \\ -2 & x = 1 \\ 3x + 5 & 1 < x \le 3 \end{cases}$$

10. _____

FORM A

10 Chapter Test (continued)   NAME

11. Determine $\lim\limits_{x\to 0} \dfrac{\cos x - 1}{x^2}$, if the limit exists.

11. _____

   A. -1 B. $-\dfrac{1}{2}$

   C. $\dfrac{1}{2}$ D. 1

   E. The limit does not exist.

12. Draw the graph of $f(x) = -2(x - 1)(x - 4)$ over the interval $[1, 4]$. On the graph, show and shade the rectangles that would be used to approximate the area under the curve $f(x)$ over $[1, 4]$ by the right rectangle approximation method using 6 subintervals. Compute the estimation.

12.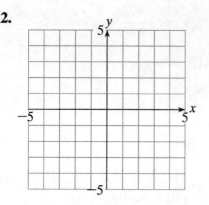

13. Write the integral that would be used to find the shaded area on the graph shown below.

13. _____

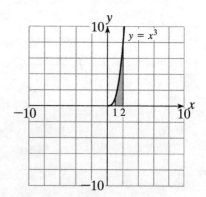

14. Explain how to find the area under the graph of $f(x) = |x - 2|$ from $x = 0$ to $x = 4$ by computing the geometric area.

14. _____
_____

15. Use a calculator to find the LRAM area approximation for the area under the graph $f(x) = -x^3 + 8$ from $x = 0$ to $x = 2$ with 20 approximating rectangles.

15. _____

FORM B

10 Chapter Test

1. Find the average rate of change of $f(x) = 5\tan x$ over the interval $\left[\dfrac{\pi}{4}, \dfrac{3\pi}{4}\right]$.

   A. $-\dfrac{5}{\pi}$ B. $-\dfrac{20}{\pi}$

   C. -10 D. 0

   1. _____

2. Find the instantaneous rate of change of $f(x) = -3x^2$ at $x = 4$.

   2. _____

3. The following table lists the population statistics for a certain city.

Year	Population
1975	995,855
1985	990,928
1995	942,547
2005	889,370

   What is the average rate of change in the population with respect to time between 1975 and 2005?

   3. _____

4. Find the derivative of $f(x) = -4x - 8$.

   4. _____

5. For the function $f(x) = \dfrac{-3}{x+2}$:

   (a) Find the slope of the function at $x = 2$ using the definition of the derivative.

   (b) What is the equation of the tangent line to the curve $f(x)$ at $x = 2$?

   5. (a) _____
   (b) _____

6. A baseball is dropped from a window. Find the baseball's (a) average speed during the first 2 sec of fall and (b) instantaneous speed at $t = 2$.

   6. (a) _____
   (b) _____

FORM B

10 Chapter Test (continued)   NAME

7. Determine the following limits, if they exist.

(a) $\lim_{x \to 4} x(x + 4)^2$

(b) $\lim_{x \to 1} \dfrac{x^2 + 3x - 4}{x^2 - 6x + 5}$

(c) $\lim_{x \to -\infty} 6x + 2$

(d) $\lim_{x \to 0} \dfrac{\sin 5x}{x}$

(e) $\lim_{x \to 4} \dfrac{x - 4}{x^2 + 2}$

(f) $\lim_{x \to 0} \dfrac{2}{x^2}$

7. (a) _____
   (b) _____
   (c) _____
   (d) _____
   (e) _____
   (f) _____

8. $\lim_{x \to 3} \dfrac{x^3 - 27}{x - 3}$

(a) Explain why direct substitution cannot be used to find the limit.

(b) Find the limit algebraically, if it exists.

8. (a) _____
   (b) _____

9. Use the graph below to find the limits or explain why the limits do not exist.

(a) $\lim_{x \to 4^-} f(x)$

(b) $\lim_{x \to 4^+} f(x)$

(c) $\lim_{x \to 4} f(x)$

9. (a) _____
   (b) _____
   (c) _____

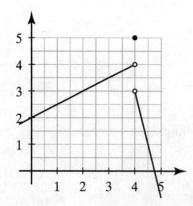

10. At what points c in the domain of $f(x)$ does $\lim_{x \to c} f(x)$ exist?

$$f(x) = \begin{cases} 3x & -5 \le x < 0 \\ 1 & x = 0 \\ x^2 + 1 & 0 < x \le 2 \end{cases}$$

10. _____

FORM B

10 Chapter Test (continued)   NAME

11. Determine $\lim_{x \to 0} \dfrac{\sin x}{x^2 - x}$, if the limit exists.

11. _____

A. -1 B. $-\dfrac{1}{2}$

C. $\dfrac{1}{2}$ D. 1

E. The limit does not exist.

12. Draw the graph of $f(x) = -x(x - 3)$ over the interval $[0, 3]$. On the graph, show and shade the rectangles that would be used to approximate the area under the curve $f(x)$ over $[0, 3]$ by the right rectangle approximation method using 6 subintervals. Compute the estimation.

12.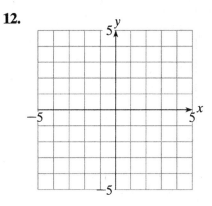

13. Write the integral that would be used to find the shaded area on the graph shown below.

13. _____

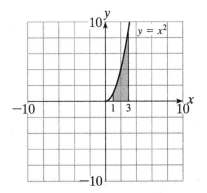

14. Explain how to find the area under the graph of $f(x) = |x - 3|$ from $x = 1$ to $x = 5$ by computing the geometric area.

14. _____

15. Use a calculator to find the RRAM area approximation for the area under the graph $f(x) = -x^3 + 8$ from $x = 0$ to $x = 2$ with 20 approximating rectangles.

15. _____

6-10 Final Exam A

1. Solve the system graphically.
   Round the answer to the nearest hundredth.

   $$\frac{x^2}{2} + \frac{y^2}{5} = 1$$

   $$y = \frac{1}{3}x$$

   1. _____

2. Solve the system by any method.

   $3x + 4y - 6z = 4$

   $x + 2y = 8$

   $5x + 2z = 4$

   2. _____

3. Which of the following is the matrix equation that represents the system of linear equations below?

   $3x + 6y = -2$

   $2x - 4y = 10$

   A. $\begin{bmatrix} 3 & 2 \\ 6 & -4 \end{bmatrix} \begin{bmatrix} x \\ y \end{bmatrix} = \begin{bmatrix} -2 \\ 10 \end{bmatrix}$

   B. $\begin{bmatrix} 3 & 6 \\ 2 & -4 \end{bmatrix} \begin{bmatrix} x \\ y \end{bmatrix} = \begin{bmatrix} 10 \\ -2 \end{bmatrix}$

   C. $\begin{bmatrix} 3 & 6 \\ 2 & -4 \end{bmatrix} \begin{bmatrix} x \\ y \end{bmatrix} = \begin{bmatrix} -2 \\ 10 \end{bmatrix}$

   D. $\begin{bmatrix} -\frac{1}{4} & -\frac{1}{2} \\ -\frac{1}{6} & \frac{1}{3} \end{bmatrix} \begin{bmatrix} x \\ y \end{bmatrix} = \begin{bmatrix} -2 \\ 10 \end{bmatrix}$

   E. $\begin{bmatrix} -\frac{1}{6} & \frac{1}{3} \\ -\frac{1}{4} & -\frac{1}{2} \end{bmatrix} \begin{bmatrix} x \\ y \end{bmatrix} = \begin{bmatrix} -2 \\ 10 \end{bmatrix}$

   3. _____

4. By augmented matrix, solve the following system:

   $x - y - z = 3$

   $3x + 4y + 5z = 1$

   $y + 2z = -2$

   4.

96

6–10 Final Exam A (continued)

5. The matrix $\begin{bmatrix} 1 & 1 & -1 \\ -3 & 2 & -1 \\ 3 & -3 & 2 \end{bmatrix}$ is the inverse of $\begin{bmatrix} 1 & 1 & 1 \\ 3 & 5 & 4 \\ 3 & 6 & 5 \end{bmatrix}$. Solve the following system by the inverse matrix method.

$x + y + z = 4$

$3x + 5y + 4z = 3$

$3x + 6y + 5z = 2$

5. _____

6. Sketch the region of feasible points and solve the linear programming problem.
Maximize $C = 4x - 5y$, subject to:

$x \geq 2$
$y \geq 0$
$x + y \leq 7$
$x + 2y \geq 8$

6. $C = $ _____

at _____

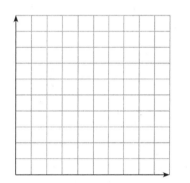

7. What is the focus of the graph of $y^2 = 16x$?

7. _____

8. Write the equation of the ellipse graphed below.

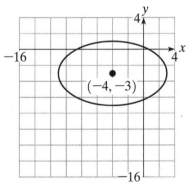

8. _____

6–10 Final Exam A (continued)

9. Use the discriminant to identify the graph of
$2x^2 - 4xy + 2y^2 + 3x - y = 5$.

   A. The graph is a circle.

   B. The graph is an ellipse.

   C. The graph is a hyperbola.

   D. The graph is a parabola.

9.

10. Find the equation that describes the set of all points (x, y, z) that are 5 units from the point $(2, -1, 6)$.

10. _____

11. What is the smallest interval for θ ($0 \le \theta \le k$) that gives a complete graph of the polar equation $r = 4 \cos 5\theta$? Sketch the graph to show which petals are generated by positive values of r and which by negative values of r.

11.

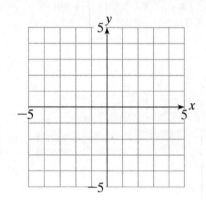

12. The graph of the polar equation $r = \dfrac{3}{4 - 3\cos\theta}$ is a(n)

   A. circle.
   B. ellipse.
   C. hyperbola.
   D. parabola opening up or down.
   E. parabola opening right or left.

12. _____

13. An airplane is flying on a bearing 23° east of north at 650 mph. Express the velocity of the airplane as a vector.

13. _____

14. At 3.5 feet above home plate, a baseball is hit with the initial velocity of 125 ft/sec at an angle of elevation of 35°. Let $t = 0$ be the time when the ball is hit. Using the position function in parametric form, find the horizontal distance the ball will travel if air resistance is neglected.

14. _____

6-10 Final Exam A (continued)

15. Determine whether the vectors $\langle 2, -1 \rangle$ and $\langle -2, -5 \rangle$ are orthogonal.

15. _____

16. Find the angle between vector $\mathbf{u} = \langle -2, 5 \rangle$ and vector $\mathbf{v} = \langle -1, 3 \rangle$.

16. _____

17. Write the parametric form of the equation of the line through the points $(2, -1, 5)$ and $(-3, 1, 0)$.

17. _____

18. Eliminate the parameter and identify the graph of the parametric curve given by $x = t - 3$, $y = \dfrac{2}{t}$.

18. _____

19. What are the four roots of $16\left(\cos \dfrac{8\pi}{15} + i \sin \dfrac{8\pi}{15}\right)$? Express the answers in trigonometric form.

19. _____

20. Use the binomial theorem to expand $(3x - 2y^2)^4$.

20. _____

21. Determine the sum, if it exists, of the infinite geometric series

$$4 + \frac{4}{3} + \frac{4}{9} + \frac{4}{27} + \ldots$$

21. _____

22. Jose deposits $200 at the end of each month into an account that pays 5% interest compounded monthly. After 7 years of deposits, what is the account balance?

22. _____

23. Write a proof by mathematical induction of the statement:

$$1^2 + 3^2 + 5^2 + \ldots + (2n - 1)^2 = \frac{n(2n - 1)(2n + 1)}{3}.$$

23. _____

24. There are 10 balls in a box that are numbered 1 through 10. A ball is selected and replaced, then a second ball is selected. What is the probability that the first ball selected will be a 3 and the second ball selected will *not* be a 9?

24. _____

6-10 Final Exam A (continued)

25. A computer randomly generates a password that is 6 entries long. The password is 3 letters followed by 3 numbers with repetitions of numbers and/or letters permitted. What is the probability that the computer will generate a password for Kara that has the same three letters followed by the same three numbers, such as GGG444?

25.

26. Find the partial fraction decomposition of $\dfrac{2x+1}{x^2}$.

26.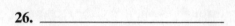

27. For the function $f(x)$ graphed below, which of the following is the $\lim\limits_{x \to 3} f(x)$?

27.

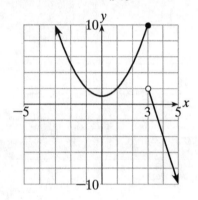

A. $\lim\limits_{x \to 3} f(x) = 10$ **B.** $\lim\limits_{x \to 3} f(x) = 3$
C. $\lim\limits_{x \to 3} f(x) = 2$ **D.** $\lim\limits_{x \to 3} f(x) = 1$
E. $\lim\limits_{x \to 3} f(x)$ does not exist

28. Find the average rate of change of $f(x) = e^x$ over the interval $[-1, 1]$.

28. _____

29. For the function $f(x) = \dfrac{1}{x}$, what is the equation of the tangent line at $x = 1$?

29. _____

30. When a car's brakes are applied, the speed of the car is given by $v(t) = 80 - 25t$ (in feet per second). Use the area under the function to determine how many feet the car travels between the time the brakes are applied ($t = 0$) and the time the car stops.

30. _____

6-10 Final Exam B

1. Solve the system graphically.
   Round the answer to the nearest hundredth.

   $$\frac{x^2}{4} + \frac{y^2}{9} = 1$$
   $$y = x^2$$

   1. _____

2. Solve the system by any method.

   $$x - 3z = 0$$
   $$x - y = 4$$
   $$x + 2y - z = 0$$

   2. _____

3. Which of the following is the matrix equation that represents the system of linear equations below?

   $$\begin{bmatrix} 4 & 7 \\ -6 & 2 \end{bmatrix} \begin{bmatrix} x \\ y \end{bmatrix} = \begin{bmatrix} 3 \\ -5 \end{bmatrix}$$

   A. $4x + 7y = 3$
      $-6x + 2y = -5$

   B. $4x - 6y = 3$
      $x + 2y = -5$

   C. $4x + 2y = -5$
      $-6x + 7y = 3$

   D. $4x + 7y = -5$
      $-6x + 2y = 3$

   E. $4x - 6y = -5$
      $7x + 2y = 3$

   3. _____

4. By augmented matrix, solve the following system:

   $$x + y + z = 1$$
   $$-2x + y + z = -2$$
   $$x - 2y + z = -2$$

   4. _____

5. The matrix $\begin{bmatrix} 1 & 2 & 1 \\ 0 & 1 & 1 \\ 1 & 0 & 1 \end{bmatrix}$ is the inverse of

   $\begin{bmatrix} 1/2 & -1 & 1/2 \\ 1/2 & 0 & -1/2 \\ -1/2 & 1 & 1/2 \end{bmatrix}$. Solve the following system by the

   inverse matrix method.

   $$x + 2y + z = 10$$
   $$y + z = 6$$
   $$x + z = -8$$

   5. _____

101

6–10 Final Exam B (continued)

6. Sketch the region of feasible points and solve the linear programming problem.
Maximize $C = -3x + 2y$, subject to:
$$x \geq 0$$
$$y \geq 3$$
$$x + 3y \leq 18$$
$$x + y \geq 8$$

6. $C =$ _____
at _____

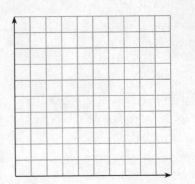

7. What are the foci for the graph of $\dfrac{x^2}{16} + \dfrac{y^2}{25} = 1$?

7.

8. Write the equation of the ellipse graphed below.

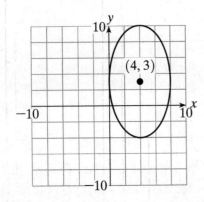

8. _____

9. Use the discriminant to identify the graph of $2x^2 + 4xy - 2y^2 - 3x - y = 8$.

A. The graph is a circle.

B. The graph is a hyperbola.

C. The graph is an ellipse.

D. The graph is a parabola.

9.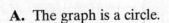

10. Find the equation that describes the set of all points (x, y, z) that are 3 units from the point $(1, -2, 6)$.

10. _____

6-10 Final Exam B (continued)

11. What is the smallest interval for θ $(0 \leq \theta \leq k)$ that gives a complete graph of the polar equation $r = 5 \sin 4\theta$? Sketch the graph to show which petals are generated by positive values of r and which by negative values of r.

11. _____

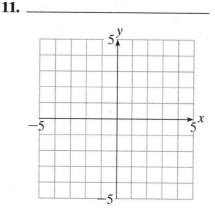

12. The graph of the polar equation $r = \dfrac{4}{3 - 4 \sin \theta}$ is a(n)

12. _____

A. circle.

B. ellipse.

C. hyperbola.

D. parabola opening up or down.

E. parabola opening right or left.

13. An airplane is flying on a bearing 75° east of north at 600 mph. Express the velocity of the airplane as a vector.

13. _____

14. A ball is thrown from the ground with initial velocity 35 ft/sec at an angle of elevation of 75°. Let $t = 0$ be the time when the ball is thrown. Using the position function in parametric form, what is the maximum height of the ball if air resistance is neglected?

14. _____

15. Determine whether the vectors $\langle 2, -1 \rangle$ and $\langle -3, -7 \rangle$ are orthogonal.

15. _____

16. Find the angle between vector $\mathbf{u} = \langle 5, -2 \rangle$ and vector $\mathbf{v} = \langle 1, -3 \rangle$.

16. _____

17. Write the parametric form of the equation of the line through the points $(-1, 2, 5)$ and $(3, -1, 0)$.

17. _____

6–10 Final Exam B (continued)

18. Eliminate the parameter and identify the graph of the parametric curve given by $x = \dfrac{3}{2t+3}, y = 2t + 2$.

18. _____

19. What are the three roots of $27\left(\cos \dfrac{9\pi}{5} + i \sin \dfrac{9\pi}{5}\right)$? Express the answers in trigonometric form.

19. _____
_____
_____

20. Use the binomial theorem to expand $(x^2 - 3y)^5$.

20. _____

21. Determine the sum, if it exists, of the infinite geometric series

$$6 - 3 + \frac{3}{2} - \frac{3}{4} + \ldots$$

21. _____

22. Pat deposits $150 at the end of each month into an account that pays 6% interest compounded monthly. After 8 years of deposits, what is the account balance?

22. _____

23. Write a proof by mathematical induction of the statement:
$1^3 + 2^3 + 3^3 + \ldots + n^3 = \dfrac{n^2(n+1)^2}{4}$.

23. _____
_____
_____

24. There are 12 balls in a box that are numbered 1 through 12. A ball is selected and replaced, then a second ball is selected. What is the probability that the first ball selected will be a 7 and the second ball selected will *not* be a 2?

24. _____

25. A computer randomly generates a password that is 8 entries long. The password is 4 letters followed by 4 numbers with repetitions of numbers and/or letters permitted. What is the probability that the computer will generate a password for Jonathan that has the same four letters followed by the same four numbers, such as KKKK5555?

25. _____

104

26. Find the partial fraction decomposition of $\dfrac{2-3x}{x^2}$.

26. _____

27. For the function $f(x)$ graphed below, which of the following is the $\lim\limits_{x \to 3} f(x)$?

27. _____

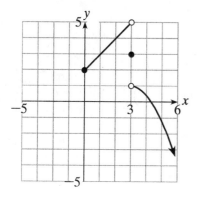

A. $\lim\limits_{x \to 3} f(x) = 2$ **B.** $\lim\limits_{x \to 3} f(x) = 1/2$
C. $\lim\limits_{x \to 3} f(x) = 1$ **D.** $\lim\limits_{x \to 3} f(x)$ does not exist

28. Find the average rate of change of $f(x) = \sin x$ over the interval $[0, \pi/2]$.

28. _____

29. For the function $f(x) = \dfrac{-2}{x}$, what is the equation of the tangent line at $x = 2$?

29. _____

30. When a car's brakes are applied, the speed of the car is given by $v(t) = 80 - 22t$ (in feet per second). Use the area under the function to determine how many feet the car travels between the time the brakes are applied ($t = 0$) and the time the car stops.

30. _____

Chapter P
Prerequisites

Quiz Sections P.1 to P.4
1. $[-2, 3)$
2. $x = 4$
3. $\sqrt{130} \approx 11.4$
4. A
5.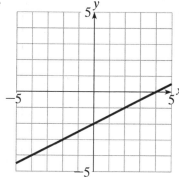
6. $\left[\dfrac{2}{3}, \infty\right)$
7. E
8. $y = \dfrac{1}{5}x - 5$
9. $\left(-\dfrac{3}{2}, 4\right), \left(\dfrac{3}{2}, 1\right), (2, 3)$
10. $\left[-\dfrac{25}{2}, \dfrac{23}{2}\right]$

Quiz Sections P.5 to P.7
1. C
2. $x = -5, x = \dfrac{3}{2}$
3. $(-3, 2)$
4. D
5. $(-\infty, -9) \cup (7, \infty)$
6. $x = -7 - 2\sqrt{10}$ and $x = -7 + 2\sqrt{10}$
7. $\dfrac{17}{2} - \dfrac{7}{2}i$
8. $(-1, 3)$
9. $x \geq 0$ or $x = -1$
10.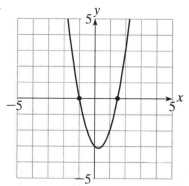

Chapter Test Form A
1. Distributive property of multiplication over addition.
2.

x	y
-2	-9
3	1
-1	-7
4	3

3. 3 years
4. $\dfrac{16}{25} + \dfrac{37}{25}i$
5. D
6. $y - 4 = 2(x - 3)$
7. $y = \dfrac{7}{2}x - 7$
8. $(x - 2)^2 + (y + 6)^2 = 81$
9. $x = \dfrac{1}{4}$ or $x = \dfrac{11}{4}$
10. $x = -0.5, x \approx -0.37,$ or $x \approx 1.37$
11. D
12. $(-\infty, -2) \cup (-1, \infty)$

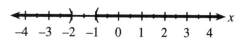

13. $\dfrac{2 \pm i\sqrt{26}}{6}$
14. $[-2, 3]$
15. $x = 4 + \sqrt{11}$ or $x = 4 - \sqrt{11}$
16. 9 seconds through 11 seconds

Chapter Test Form B
1. Commutative property of addition
2.

x	y
-2	8
3	-7
2	-4
-1	5

3. 4 years
4. $\dfrac{7}{10} - \dfrac{8}{5}i$
5. C
6. $y - 3 = -4(x - 1)$
7. $y = \dfrac{5}{3}x - 5$
8. $(x + 3)^2 + (y - 4)^2 = 49$
9. $x = -2$ or $x = \dfrac{4}{5}$
10. $x \approx -1.62, x \approx 0.33,$ or $x \approx 0.62$

11. D

12. $\left(-\infty, \frac{2}{3}\right] \cup [2, \infty)$;

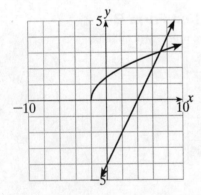

13. $\dfrac{-3 \pm i\sqrt{19}}{4}$

14. $[-4, 1]$

15. $x = \dfrac{9}{2} - \dfrac{1}{2}\sqrt{65}$, $x = \dfrac{9}{2} + \dfrac{1}{2}\sqrt{65}$

16. 6 seconds through 16 seconds

Chapter 1
Functions and Graphs

Quiz Sections 1.1 to 1.3
1. $[5, \infty]$
2. $[8, \infty]$
3. $(-3, -5)$
4. Answers will vary. One answer is $y = \dfrac{1}{x}$; discontinuous at $x = 0$.
5. 180,000 sq. ft.
6. $x = 2.8$, $x = -2.9$
7. Solution: $x = 7$
   Extraneous solution: $x = 2$

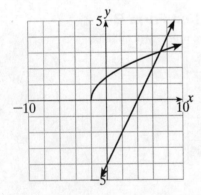

8. (a) $0, -\sqrt{2}, \sqrt{2}$
   (b) increasing on $(-1, 0)$ and $(1, \infty)$
9. C
10. E

Quiz Sections 1.4 to 1.7
1. B
2.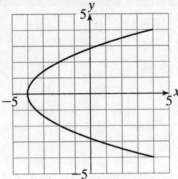

3. Horizontal shrink of factor $\dfrac{1}{2}$.
4. $g(x) = -2x^3 - 3x^2 - 5x + 6$
5. $y = -4x^2 + 5$
6. $f^{-1}(x) = \dfrac{x + 5}{3x - 6}$
7. $y = -x + 2$ and $y = x - 2$
   Domain: $x \leq 2$
8. 22%
9. $0.12x + 0.20(20) = 0.15(x + 20)$; $33\dfrac{1}{3}$ gallons
10. (a) $A = 1.23x$
    (b) 37.40

Chapter Test Form A
1. C
2. Vertical: $x = 5$, $x = -2$; Horizontal: $y = 0$
3. (a)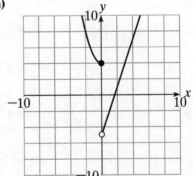

(b) Not continuous; a jump occurs at $x = 0$.

4. $x = 2$, $x = -5$
5. 57 mph
6. $\left(-\infty, \dfrac{3}{2}\right) \cup \left(\dfrac{5}{2}, \infty\right)$

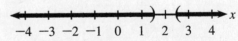

Answers

7. $f(g(x)) = \dfrac{12}{x^2} - 5$; domain: $\{x|x \neq 0\}$

8. $V = 13{,}500\pi - 5t$

9.
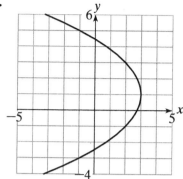

10. Shift the graph of $y = x^2$ one unit to the left, then stretch vertically by a factor of 4.

11. $y = x^2,\ y = \cos(x),\ y = |x|$

12. $x = t,\ y = -16t^2 + 60t + 3$
Max. ht. $= 59.25$ ft
Time $= 1.88$ sec

13. Min at $x \approx 2.59$
Max at $x \approx -0.26$

14. (a) f passes the horizontal and vertical line tests and hence is one-to-one.
(b) $f^{-1}(x) = 2 - x^2$, domain of $f^{-1}(x):[0, \infty)$

15. Upper: $y = 3$; Lower: $y = 1$

■ Chapter Test Form B

1. B

2. Vertical: $x = -\dfrac{1}{2},\ x = 6$; Horizontal: $y = 0$

3. (a)
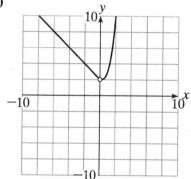

(b) Discontinuous, hole at $x = 0$

4. $x = 3,\ x = -4$

5. 34 mph

6. $\left(-\infty, -\dfrac{7}{3}\right) \cup \left(-\dfrac{5}{3}, \infty\right)$

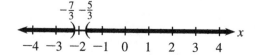

7. $f \circ g = -\dfrac{5x^2}{4}$; Domain: $(-\infty, \infty)$

8. $V(t) = 6{,}250\pi - 4t$

9.
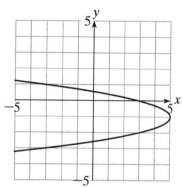

10. Shift the graph of $y = x^2$ one unit to the right, then stretch vertically by a factor of 5.

11. $y = x,\ y = x^3,\ y = \dfrac{1}{x},\ y = \sin(x),\ y = \tan(x)$

12. $x = t,\ y = -16t^2 + 65t + 4$
Max. ht. $= 70.02$ ft
Time $= 2.03$ sec

13. Min at $x = 2$
Max at $x = -0.33$

14. (a) f passes the horizontal and vertical line tests and hence is one-to-one.
(b) $f^{-1}(x) = x^2 + 5$, Domain of $f^{-1}(x):[0, \infty)$

15. Upper: $y = 1$
Lower: $y = -2$

Chapter 2
Polynomial, Power, and
Rational Functions

■ Quiz Sections 2.1 to 2.4

1. $y = 4x^2 - 24x + 34$

2. B

3. $f(x) \to \underline{\ \infty\ }$ as $x \to -\infty$;
$f(x) \to \underline{\ -\infty\ }$ as $x \to \infty$

4. 28   **5.** D

6. $5x - 1$

7. $3x^2 - 8x + 21 - \dfrac{45}{x+2}$

8. $V = -3{,}000t + 45{,}000$

9. Maximum height: 81 ft
Time: 2.25 sec

10. (a) $R(x) = -25x^2 + 1{,}900x + 60{,}000$
(b) $\$1{,}550$ rent per month
(c) $\$96{,}100$

Quiz Sections 2.5 to 2.8

1. D
2. $-\dfrac{5}{2}, -1, 4$
3. $-1, 2, i, -i$
4. $(x - 1)\left(x + \dfrac{1}{2} - \dfrac{3\sqrt{3}}{2}i\right)\left(x + \dfrac{1}{2} + \dfrac{3\sqrt{3}}{2}i\right)$
5. D
6. $(-\infty, -5]$
7. Root: $\dfrac{-15}{4}$
   Extraneous root: 2
8. $f(x) = x^2 - 6x + 13$
9. $(-\infty, -1) \cup \left[-\dfrac{2}{3}, 0\right)$
10. y-intercept: $\dfrac{5}{2}$; vertical asymptote: $x = -2$; slant asymptote: $y = x - 5$

Chapter Test Form A

1. Quotient: $x^2 + x + 7$;
   Remainder: 19
2. 2
3. $1,600
4. D
5. Zeros: $3, -1 \pm \sqrt{6}\,i$;
   $f(x) = (x - 3)(x + 1 - \sqrt{6}\,i)(x + 1 + \sqrt{6}\,i)$
6. 128
7. $y = x^2 - 6x + 5$
8. 10 in. × 14 in.
9. Possible answer: (see graph)

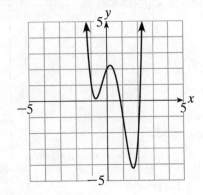

10. $-\infty; -\infty$
11. Horizontal: $y = 3$
    Vertical: $x = 3$ and $x = 4$
12. $(-\infty, 2) \cup (2, 6]$

13. $D = 0.266t^3 - 1.955t^2 + 3.118t + 2.811$

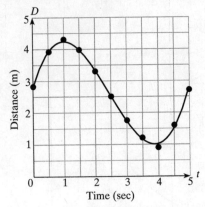

14. Asymptotes:
    $x = -3, x = 2$,
    $y = 0$;
    Intercepts:
    $(5, 0), \left(0, \dfrac{5}{6}\right)$

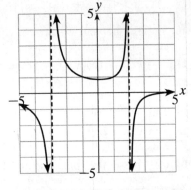

15. Translate 2 units left, stretch vertically by a factor of 4, and translate 3 units down. The order may be changed provided the vertical stretch precedes the vertical translation.
16. $(-2, 0) \cup [5, \infty)$

Chapter Test Form B

1. Quotient: $x^2 + 5x + 2$;
   Remainder: 11
2. 18
3. $1,130
4. C
5. Zeros: $-2, -1 \pm \sqrt{10}\,i$.
   $f(x) = (x - 2)(x + 1 - \sqrt{10}\,i)(x + 1 + \sqrt{10}\,i)$
6. -106
7. $y = x^2 - 6x + 2$
8. 24 ft. × 32 ft.
9. Possible answer: (see graph)

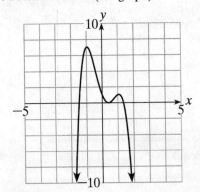

Answers 111

10. ∞; $-\infty$

11. Horizontal: $y = \dfrac{5}{2}$

    Vertical: $x = 4$ and $x = \dfrac{3}{2}$

12. $[4, \infty)$

13. $D = -0.266t^3 + 1.990t^2 - 3.193t + 2.206$

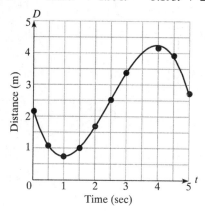

14. Asymptotes: $x = -4$, $x = 3$, $y = 0$;

    Intercepts: $(-6, 0)$, $\left(0, -\dfrac{1}{2}\right)$

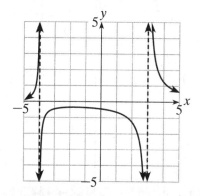

15. Translate 4 units right, stretch vertically by a factor of 2, and translate 5 units up. The order may be changed provided the vertical stretch precedes the vertical translation.

16. $(-\infty, -3) \cup (0, 4]$

Chapter 3
Exponential, Logistic, and Logarithmic Functions

■ Quiz Sections 3.1 to 3.4

1. $g(x) \to$ _____ ∞ _____ as $x \to -\infty$; $g(x) \to$ _____ 0 _____ as $x \to \infty$

2. E

3. 13.81 years

4.

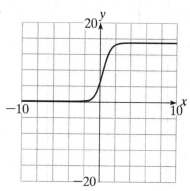

5. $\log_3 9 = x$
6. $f^{-1}(x) = \ln x - 2$
7. Vertical stretch by a factor of 5 to obtain $y = 5 \ln x$, horizontal shift 1 unit right to obtain $y = 5 \ln (x - 1)$, vertical shift 3 units up to obtain $y = 5 \ln (x - 1) + 3$.
8. C
9. $x = \sqrt[3]{5}$
10. 1.64

■ Quiz Sections 3.5 to 3.6

1. A
2. $x = 2$; Extraneous $x = -4$
3. B
4. $x = -1.45, 0$
5. 42.58 minutes
6. $8346.05
7. 6.18%
8. $28,649.40
9. 3.32 years
10. $874.02

■ Chapter Test Form A

1.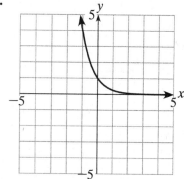

2. (a) 3 (b) 9
3. About 53 days
4. Equation: $y = -0.50 + 5.69 \ln x$; y-value: 14.91

5. $f(g(x)) = f(e^{4x}) = \frac{1}{4}\ln(e^{4x}) = \frac{1}{4}(4x) = x$

$g(f(x)) = g\left(\frac{1}{4}\ln x\right) = e^{4(1/4 \ln x)} = e^{\ln x} = x$

6. (a) $y = 0, y = 5$

(b) Domain $(-\infty, \infty)$ Range $(0, 5)$

(c) $\lim_{x \to \infty} f(x) = 5$ $\lim_{x \to -\infty} f(x) = 0$

7. Translate 2 units left and 1 unit up (in either order).

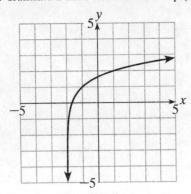

8. $x = e^{-4/3} \approx 0.26$

9. Solution: $x = 7$; Extraneous: $x = 0$; $x = 0$ is extraneous because $\log(0-5)$ and $\log(0-2)$ are undefined.

10. $x = \frac{5}{3}$

11. About 36.81 min

12. A

13. $23,369.26

14. $3,853.73

15. $115,510.22

16. D

■ Chapter Test Form B

1.

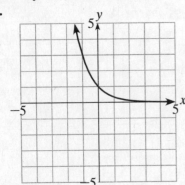

2. (a) 4 **(b)** 5

3. About 40 days

4. Equation: $y = -1.42 + 5.10 \ln x$; y-value: 13.32

5. $f(g(x)) = f(e^{5x}) = \frac{1}{5}\ln(e^{5x}) = \frac{1}{5}(5x) = x$

$g(f(x)) = g\left(\frac{1}{5}\ln x\right) = e^{5(1/5 \ln x)} = e^{\ln x} = x$

6. (a) $y = 0, y = 6$

(b) Domain: $(-\infty, \infty)$

(c) $\lim_{x \to \infty} f(x) = 6$, $\lim_{x \to -\infty} f(x) = 0$

7. Translate 4 units left and reflect across the x-axis (in either order).

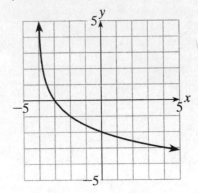

8. $x = 10^{-3/4} \approx 0.18$

9. Solution: $x = 5$; Extraneous: $x = -6$; $x = -6$ is extraneous because $\log(-6-4)$ and $\log(-6+5)$ are undefined.

10. $x = 2$

11. About 65.03 minutes

12. D

13. $32,925.80

14. $4,492.38

15. $366,148.70

16. C

Chapter 4
Trigonometric Functions

■ Quiz Sections 4.1 to 4.4

1. $\sin \theta = \frac{5\sqrt{34}}{34}$, $\cos \theta = -\frac{3\sqrt{34}}{34}$, $\tan \theta = -\frac{5}{3}$,

$\csc \theta = \frac{\sqrt{34}}{5}$, $\sec \theta = -\frac{\sqrt{34}}{3}$, $\cot \theta = -\frac{3}{5}$

2. A

3. $\frac{-2\sqrt{21}}{21}$

4. 1.73 yards

5. 0.6248

6. $\frac{\pi}{3}$

7. B

Answers 113

8. 2
9. 57.98°
10.
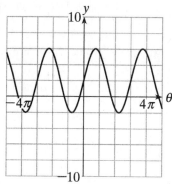

■ Quiz Sections 4.5 to 4.8

1. $\dfrac{\pi}{2}$
2. B
3.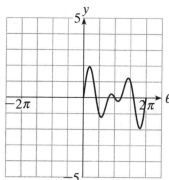
4. Domain: $(-\infty, \infty)$
   Range: $[0, 1]$
   Period: π
5.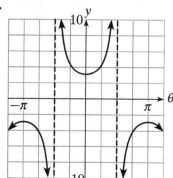
6. 3.61
7. A
8. 109.19 feet
9. 423.02 feet
10. 14,600 units

■ Chapter Test Form A

1. $\sin\theta = \dfrac{15}{17}, \cos\theta = \dfrac{8}{17}, \tan\theta = \dfrac{15}{8}, \csc\theta = \dfrac{17}{15},$
   $\sec\theta = \dfrac{17}{8}, \cot\theta = \dfrac{8}{15}$
2. $\sec 2 \approx -2.40$
3. 16 inches
4. $\alpha = 52°; a \approx 5.76; c \approx 7.31$
5. B
6. Amplitude: 6; Period: $\dfrac{2\pi}{3}$; Phase shift: $\dfrac{\pi}{12}$; Vertical translation: 2
7. (a) No
   (b) Yes, 3^{-x}
   (c) Yes, $-4x$
   (d) No
   (e) Yes, $0.35e^{-0.07x}$
8. The graph of $y = 3\csc\dfrac{1}{2}x$ can be obtained from the graph of $y = \csc x$ by a vertical stretch of factor 3 and a horizontal stretch of factor 2. Period: 4π; Domain: all reals except $x = 2\pi n$, n any integer; Range: $(-\infty, -3] \cup [3, \infty)$; Zeros: none; Asymptotes: $x = n2\pi$, n any integer

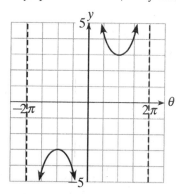

9. $x \approx 3.50$ or $x \approx 6.64$
10. $a \approx 4.47; b = 5.00; h \approx 0.22$
11. B
12. 327.4 meters
13. $\cos(\sin^{-1}u) = \sqrt{1-u^2}$; The $\pm$ sign is unnecessary because the range of $\sin^{-1}u$ is $\left[-\dfrac{\pi}{2}, \dfrac{\pi}{2}\right]$ and $\cos x \geq 0$ for $x \in \left[-\dfrac{\pi}{2}, \dfrac{\pi}{2}\right]$.
14. (a) $80 million
    (b) $92.5 million
    (c) 8 years

114 Tests and Quizzes

15. Period: 2π

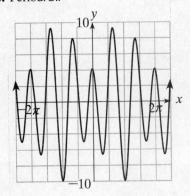

■ Chapter Test Form B

1. $\sin\theta = \dfrac{12}{13}$; $\cos\theta = \dfrac{5}{13}$; $\tan\theta = \dfrac{12}{5}$; $\csc\theta = \dfrac{13}{12}$;
   $\sec\theta = \dfrac{13}{5}$; $\cot\theta = \dfrac{5}{12}$

2. $\csc 3 \approx 7.09$

3. 16.8 inches

4. $\beta = 47°$; $b \approx 3.75$; $c \approx 5.13$

5. C

6. Amplitude: 2; Period: $\dfrac{\pi}{2}$; Phase shift: $\dfrac{\pi}{20}$; Vertical translation: -3

7. (a) Yes, 4^{-x}
   (b) No
   (c) Yes, $-5x$
   (d) No
   (e) Yes, $45e^{-0.1x}$

8. The graph of $y = -\sec 2x$ can be obtained from the graph of $y = \sec x$ by a horizontal shrink of factor $\dfrac{1}{2}$ and a reflection across the x-axis. Period: π; Domain: $x \neq \dfrac{\pi}{4} + \dfrac{n\pi}{2}$, n any integer; Range: $(-\infty, -1] \cup [1, \infty)$; Zeros: none; Asymptotes: $x = \dfrac{\pi}{4} + \dfrac{n\pi}{2}$, n any integer

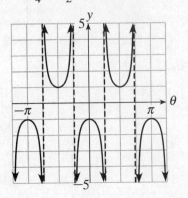

9. $x \approx 5.98$ or $x \approx 9.12$

10. $a \approx 7.21$; $b = 3.00$; $h \approx 0.33$

11. E

12. 273.6 meters

13. $\csc(\cos^{-1} u) = \dfrac{1}{\sqrt{1-u^2}}$; The $\pm$ sign is unnecessary because the range of $\cos^{-1} u$ is $[0, \pi]$ and $\csc x \geq 0$ for $x \in [0, \pi]$.

14. (a) $68 million
    (b) $125.8 million
    (c) 6 years

15. Period: 2π

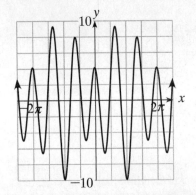

Chapter 5
Analytic Geometry

■ Quiz Sections 5.1 to 5.3

1. $\cot\theta$

2. $1 + \csc x$

3. B

4. $\dfrac{\sin\theta}{1 + \cos\theta} + \dfrac{1 + \cos\theta}{\sin\theta} = \dfrac{\sin^2\theta + (1 + \cos\theta)^2}{\sin\theta(1 + \cos\theta)}$

$= \dfrac{\sin^2\theta + 1 + 2\cos\theta + \cos^2\theta}{\sin\theta(1 + \cos\theta)}$

$= \dfrac{(\sin^2\theta + \cos^2\theta) + 1 + 2\cos\theta}{\sin\theta(1 + \cos\theta)}$

$= \dfrac{2 + 2\cos\theta}{\sin\theta(1 + \cos\theta)}$

$= \dfrac{2(1 + \cos\theta)}{\sin\theta(1 + \cos\theta)}$

$= \dfrac{2}{\sin\theta} = 2\csc\theta$

5. $\cos\theta + \cos 2\theta + \cos(2\theta + \theta)$
$= \cos\theta + \cos 2\theta + \cos 2\theta \cos\theta - \sin 2\theta \sin\theta$
$= \cos\theta + \cos 2\theta + \cos 2\theta \cos\theta - 2\sin^2\theta \cos\theta$
$= \cos\theta + \cos 2\theta + \cos 2\theta \cos\theta - 2\cos\theta(1 - \cos^2\theta)$
$= \cos 2\theta + \cos 2\theta \cos\theta - \cos\theta + 2\cos^3\theta$
$= \cos 2\theta + \cos 2\theta \cos\theta + \cos\theta(2\cos^2\theta - 1)$
$= \cos 2\theta + \cos 2\theta \cos\theta + \cos\theta \cos 2\theta$
$= \cos 2\theta + 2\cos 2\theta \cos\theta$
$= \cos 2\theta(1 + 2\cos\theta)$

6. D

7. $x = \dfrac{\pi}{2} + n\pi$, $x = \dfrac{\pi}{4} + \dfrac{n\pi}{2}$, where n is any integer.

8. $x = \dfrac{\pi}{6}$, $x = \dfrac{5\pi}{6}$

9. $\left[\dfrac{\pi}{6}, \dfrac{5\pi}{6}\right] \cup \left[\dfrac{7\pi}{6}, \dfrac{11\pi}{6}\right]$

10. This is not an identity. For example, if $x = 0$
$2\cos(0) = 2 \neq \sin(2 \cdot 0) = 0$.

■ Quiz Sections 5.4 to 5.6

1. $\gamma = 113°$, $a = 9.50$, and $c = 13.07$
2. C
3. 6,040 meters
4. 36.98 sq units
5. 142.19 sq units
6. 26.0 meters
7. C
8. $\dfrac{\sqrt{2 - \sqrt{2}}}{2}$
9. 63.29 ft
10. $\dfrac{\sin^3 A - \cos^3 A}{\sin^2 A - \cos^2 A}$
$= \dfrac{(\sin A - \cos A)(\sin^2 A + \sin A \cos A + \cos^2 A)}{(\sin A + \cos A)(\sin A - \cos A)}$
$= \dfrac{(1 + \sin A \cos A)}{(\sin A + \cos A)}$
$= \dfrac{1 + \frac{1}{2}\sin 2A}{\sin A + \cos A}$
$= \dfrac{2 + \sin 2A}{2(\sin A + \cos A)}$

■ Chapter Test Form A

1. $\sec^2 x - \dfrac{\sec^2 x}{\csc^2 x} = \dfrac{1}{\cos^2 x} - \dfrac{\frac{1}{\cos^2 x}}{\frac{1}{\sin^2 x}} = \dfrac{1 - \sin^2 x}{\cos^2 x}$
$= \dfrac{\cos^2 x}{\cos^2 x} = 1.$

2. $(\tan x + 1)^2 = \tan^2 x + 2\tan x + 1$
$= (\tan^2 x + 1) + 2\tan x$
$= \sec^2 x + 2\tan x$
$= \dfrac{1}{\cos^2 x} + 2\dfrac{\sin x}{\cos x} = \dfrac{1 + 2\sin x \cos x}{\cos^2 x}$

3. $x = -\dfrac{\pi}{2} + 2\pi n$, or $x = \dfrac{\pi}{4} + n\pi$ where n is any integer

4. No; any value of x (except $\dfrac{\pi}{2} + n\pi$, n an integer) is a counter example.

5. $\dfrac{\sqrt{6} - \sqrt{2}}{4}$

6. $\cos 3x = \cos x \cos 2x - \sin x \sin 2x$
$= \cos x (\cos^2 x - \sin^2 x) - \sin x(2\sin x \cos x)$
$= \cos^3 x - 3\sin^2 x \cos x$

7. $\pm\sqrt{\dfrac{1 - \cos 6C}{2}}$

8. $\beta = 59.16°$

9. 0

10. $\theta \approx 25.41°$ or $\theta \approx 64.59°$

11. 52,440 square feet

12. 12.2 meters

13. About 646.7 ft

14. 87.5 meters

15. $A(\theta) = 64\cos\dfrac{\theta}{2}\sin\dfrac{\theta}{2} = 32\sin\theta$
$\theta \approx 51.38°$ or $\theta \approx 128.62°$

■ Chapter Test Form B

1. $\tan^2 x - \dfrac{\csc^2 x}{\cot^2 x} = \dfrac{\sin^2 x}{\cos^2 x} - \dfrac{\frac{1}{\sin^2 x}}{\frac{\sin^2 x}{\cos^2 x}} \cdot \dfrac{\sin^2 x}{\cos^2 x} = \dfrac{\sin^2 x - 1}{\cos^2 x}$

$= \dfrac{-\cos^2 x}{\cos^2 x} = -1.$

2. $(1 + \cot x)^2 = 1 + 2\cot x + \cot^2 x$
$= (1 + \cot^2 x) + 2\cot x$
$= \csc^2 x + 2\cot x$
$= \dfrac{1}{\sin^2 x} + 2\dfrac{\cos x}{\sin x} = \dfrac{1 + 2\sin x \cos x}{\sin^2 x}$

3. $x = n\pi$, or $x = \dfrac{\pi}{2} + 2n\pi$ where n is any integer

4. No; any value of x (except $x = n\pi$, n an integer) is a counter example.

5. $\dfrac{-(\sqrt{6} + \sqrt{2})}{4}$

6. $\sin 3x = \sin x \cos 2x + \cos x \sin 2x$
$= \sin x (2\cos^2 x - 1) + \cos x(2\sin x \cos x)$
$= 4\sin x \cos^2 x - \sin x$

7. $\pm\sqrt{\dfrac{1 + \cos 8C}{2}}$

8. 53.67°

9. 1

10. $\theta \approx 19.90°$ or $\theta \approx 70.10°$

11. 24,495 square feet

12. 12.9 meters

13. Approximately 616.9 feet
14. 216.4 meters
15. $A(\theta) = 100 \cos \frac{\theta}{2} \sin \frac{\theta}{2} = 50 \sin \theta$
    $\theta \approx 47.73°$ or $\theta \approx 132.27°$

Chapters P–5
Midterm Exam

■ Midterm Exam A

1. $\left(-\frac{18}{5}, -\frac{8}{5}\right]$

2. Sample answer:
$AB = \sqrt{[1 - (-8)]^2 + (-3 - 1)^2}$
$= \sqrt{9^2 + (-4)^2} = \sqrt{81 + 16} = \sqrt{97};$
$BC = \sqrt{[-8 - (-4)]^2 + (1 - 10)^2}$
$= \sqrt{(-4)^2 + (-9)^2} = \sqrt{16 + 81} = \sqrt{97};$
$CD = \sqrt{(-4 - 5)^2 + (10 - 6)^2}$
$= \sqrt{(-9)^2 + 4^2} = \sqrt{81 + 16} = \sqrt{97};$
$AD = \sqrt{(1 - 5)^2 + (-3 - 6)^2}$
$= \sqrt{(-4)^2 + (-9)^2} = \sqrt{16 + 81} = \sqrt{97};$
$m_{\overline{AB}} = \frac{1 - (-3)}{-8 - 1} = -\frac{4}{9}$ and
$m_{\overline{BC}} = \frac{10 - 1}{-4 - (-8)} = \frac{9}{4},$
so $\overline{AB} \perp \overline{BC}$ since $m_{\overline{AB}} \cdot m_{\overline{BC}} = -1$. Since all four sides have the same length and since one pair of adjacent sides are perpendicular, the points are the vertices of a square. (Note: Other proofs are possible; check students' work.)

3. $\sqrt[35]{x^3}$

4. $y = -\frac{2}{5}x + \frac{31}{5}$ or $y = -0.4x + 6.2$

5. $y = 256.51x + 15,326.48;$
2008: About $23,791

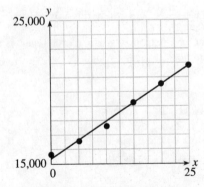

6. Approximately $(-\infty, 29.43]$

7. Solution: $x = 8$; Extraneous root: $x = 1$. The root $x = 1$ is extraneous because the domain of the logarithm function is the positive real numbers.

8. $x \approx -1.30$ or $x \approx 4.37$
9. $g \circ f(x) = x - 2$; Domain: $[3, \infty)$
10. $A(x) = 240x - 2x^2$; $x \approx 30.85$ or $x \approx 89.15$
11. $(-\infty, 4) \cup (12, \infty)$
12. Center: $(-10, 8)$; $r = 2\sqrt{21}$
13. $y = 3x^2 - 12x + 7$
14.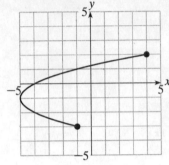

15. $y = 3(x - 5)^2 - 6$
16. B
17. x-intercepts: $(4, 0), (-4, 0)$
    y-intercept: $\left(0, \frac{32}{9}\right)$
    Vertical asymptotes: $x = 3, x = -3$
    Horizontal asymptote: $y = 2$
18. Sample answer: According to the lower bound test for real zeros, -2 is a lower bound for the zeroes of $f(x)$ if and only if 2 is an upper bound for the zeros of $f(-x) = x^4 + 3x^3 - 4x^2 - 8x - 2$. Applying the upper bound test for real zeros, we obtain the synthetic division

    $\begin{array}{r|rrrrr} 2 & 1 & 3 & -4 & -8 & -2 \\ & & 2 & 10 & 12 & 8 \\ \hline & 1 & 5 & 6 & 4 & 6 \end{array}$

    Since the last row contains no negative numbers, 2 is an upper bound for the zeros of $f(-x)$ and -2 is a lower bound for the zeros of $f(x)$.

19. $x = 2 \pm \sqrt{17}i$
20. $x^3 + x^2 - 32x + 70$
21. $e^{(x/2)} - 3$
22. C
23. Translate 2 units left, stretch vertically by a factor of 4, and translate 3 units down. The order may be changed as long as the vertical stretch precedes the downward translation.
24. $13,374.18
25. After 6 years
26. (a) 16
    (b) After 159.78 years
    (c) 1216
27. $\sin \theta = 15/17$ $\cos \theta = 8/17$
    $\tan \theta = 15/8$ $\sec \theta = 17/8$
    $\csc \theta = 17/15$ $\cot \theta = 8/15$

28. $\alpha = 52°$ $a \approx 5.76$ $c \approx 7.31$
29. B
30. 64.6 feet
31. B
32. (a) $108 million
    (b) $117.5 million
    (c) 6 years

Midterm Exam B

1. $[-4, 2)$
2. Sample answer:
$$AB = \sqrt{[4-(-1)]^2 + (2-0)^2}$$
$$= \sqrt{5^2 + 2^2} = \sqrt{25+4} + \sqrt{29};$$
$$BC = \sqrt{[-1-(-3)]^2 + (0-5)^2}$$
$$= \sqrt{2^2 + (-5)^2} = \sqrt{4+25} = \sqrt{29};$$
$$CD = \sqrt{[2-(-3)]^2 + (7-5)^2}$$
$$= \sqrt{5^2 + 2^2} = \sqrt{25+4} = \sqrt{29};$$
$$AD = \sqrt{(4-2)^2 + (2-7)^2}$$
$$= \sqrt{2^2 + (-5)^2} = \sqrt{4+25} = \sqrt{29};$$
$$m_{\overline{AB}} = \frac{0-2}{-1-4} = \frac{2}{5} \text{ and } m_{\overline{BC}} = \frac{5-0}{-3-(-1)} = -\frac{5}{2},$$
so $\overline{AB} \perp \overline{BC}$ since $m_{\overline{AB}} \cdot m_{\overline{BC}} = -1$. Since all four sides have the same length and since one pair of adjacent sides are perpendicular, the points are the vertices of a square. (Note: Other proofs are possible; check students' work.)
3. $\sqrt[15]{x}$
4. $y = \frac{3}{8}x + \frac{41}{8}$ or $y = 0.375x + 5.125$
5. $y = 316.89x + 18{,}295.19$;
   2012: About $30,020

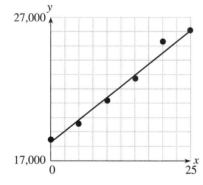

6. Approximately $(-\infty, 15.25]$
7. Solution: $x = 5$; Extraneous root: $x = -6$. The root $x = -6$ is extraneous because the domain of the logarithm function is the positive real numbers.
8. $x \approx -0.71$ or $x \approx 3.31$
9. $f \circ g(x) = x + 4$; Domain: $[-5, \infty)$
10. $A(x) = 160x - 2x^2$; $x \approx 25.86$ or $x \approx 54.14$
11. $(-\infty, 2] \cup [8, \infty)$
12. Center: $(12, -6)$; $r = 4\sqrt{5}$
13. $y = 2x^2 + 12x + 22$
14.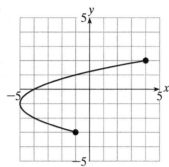

15. $y = 0.3(x+5)^2 + 6$
16. C
17. x-intercepts: $(5, 0), (-5, 0)$
    y-intercept: $\left(0, \frac{75}{4}\right)$
    Vertical asymptotes: $x = 2, x = -2$
    Horizontal asymptote: $y = 3$
18. Sample answer: According to the lower bound test for real zeros, -3 is a lower bound for the zeroes of $f(x)$ if and only if 3 is an upper bound for the zeros of $f(-x) = x^4 + 2x^3 - 8x^2 - 3x - 10$. Applying the upper bound test for real zeros, we obtain the synthetic division

3 |	1	2	−8	−3	−10
		3	15	21	54
	1	5	7	18 |	44

    Since the last row contains no negative numbers, 3 is an upper bound for the zeros of $f(-x)$ and -3 is a lower bound for the zeros of $f(x)$.
19. $x = 3 \pm 7i$
20. $x^3 - 21x^2 + 148x - 340$
21. $y = e^{(x-4)} - 1$
22. D
23. Translate 4 units right, reflect across the x-axis, and translate 5 units up. The order may be changed as long as the reflection precedes the upward translation.
24. $10,932.03
25. After 2 years and 7 months
26. (a) 23
    (b) After 88.31 years
    (c) 1518
27. $\sin \theta = 12/13$ $\cos \theta = 5/13$
    $\tan \theta = 12/5$ $\sec \theta = 13/5$
    $\csc \theta = 13/12$ $\cot \theta = 5/12$
28. $\beta = 47°$ $b \approx 3.75$ $c \approx 5.13$
29. B

Chapter 6
Applications of Trigonometry

Quiz Sections 6.1 to 6.3

1. $\frac{3}{5}\mathbf{i} - \frac{4}{5}\mathbf{j}$
2. D
3. $(315.47, 450.53)$
4. $176.63°$
5. $\langle 5 \cos(242°), 5 \sin(242°) \rangle = \langle -2.35, -4.41 \rangle$
6. $x = y^2 - 10y + 27$
7. $x(t) = 5 + 3t$
   $y(t) = -4 - 10t$
8. $\langle 3, -3 \rangle$
9. C
10. 4.55 seconds

Quiz Sections 6.4 to 6.6

1. $\left(4\sqrt{2}, \frac{3\pi}{4}\right), \left(-4\sqrt{2}, \frac{7\pi}{4}\right)$
2. D
3. $x^2 - 3y + y^2 = 0$
4. $\sqrt{13}$;

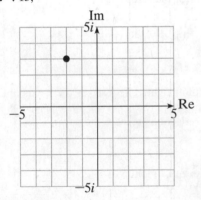

5. $-\sqrt{3} + 3i$
6. $r = 4 \cos 2\theta$

7.

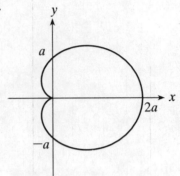

8.

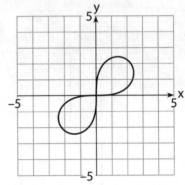

9. B
10. $\cos 2\pi/5 + i \sin 2\pi/5$, $\cos 4\pi/5 + i \sin 4\pi/5$,
    $\cos 6\pi/5 + i \sin 6\pi/5$, $\cos 8\pi/5 + i \sin 8\pi/5$,
    $\cos 0 + i \sin 0 = 1 + 0i$

Chapter Test Form A

1. $-5\mathbf{i} + \mathbf{j}$
2. $\langle -10, 19 \rangle$
3. Ground speed ≈ 437.90 mph; Bearing $\approx 261.78°$ or $8.22°$ south of due west.
4. $\langle 7/25, -24/25 \rangle = \langle 0.28, -0.96 \rangle$
5. $\langle 2, -2 \rangle$
6. $12(\cos 3\pi/4 + i \sin 3\pi/4)$
7. $\langle 3, 3 \rangle$
8. Approximately $(6.55, -4.59)$
9. $\frac{7}{4} + i\frac{7\sqrt{3}}{4}$
10. parabola; $y = 1 - x^2$
11. (a) $x(t) = (90 \cos 70°)t$;
       $y(t) = -16t^2 + (90 \sin 70°)t$
    (b) $0 \le t \le 5.29$
    (c) ≈ 163 ft
12. A
13. About 7.63 miles

14. $0 \leq \theta \leq \pi$;

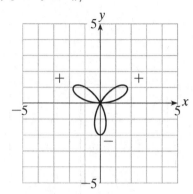

15. $5(\cos \theta + i \sin \theta)$, where $\theta = 5\pi/12, 3\pi/4, 13\pi/12,$ $17\pi/12, 7\pi/4$

16. $15{,}625i$

Chapter Test Form B
1. $-7\mathbf{i} + 2\mathbf{j}$
2. $\langle -13, 14 \rangle$
3. Ground speed ≈ 418.36 mph; Bearing $\approx 143.67°$ or 36.33° east of due south.
4. $\langle -4/5, 3/5 \rangle = \langle -0.8, 0.6 \rangle$
5. $\langle 5, -5 \rangle$
6. $6(\cos 2\pi/3 + i \sin 2\pi/3)$
7. $\langle 4, 4 \rangle$
8. Approximately $(-6.66, 2.16)$
9. $-\dfrac{9}{8}\sqrt{2} + \dfrac{9}{8}\sqrt{2}i$
10. hyperbola; $y = 1 + \dfrac{2}{x}$
11. (a) $x(t) = (95 \cos 65°) t$; $y(t) = -16t^2 + (95 \sin 65°)t$
    (b) $0 \leq t \leq 5.38$
    (c) ≈ 216 ft
12. A
13. About 608 miles
14. $0 \leq \theta \leq 2\pi$;

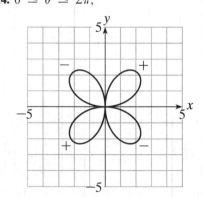

15. $6(\cos \theta + i \sin \theta)$, where $\theta = \pi/2, 9\pi/10, 13\pi/10,$ $17\pi/10$

16. $7776i$

Chapter 7
Systems and Matrices

Quiz Sections 7.1 to 7.2
1. E
2. $(4, -3)$
3. $(2, 3), (-2, 3)$
4. $(4, 2), (-4, -2)$

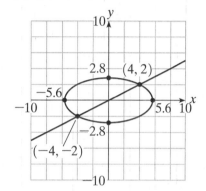

5. -20

6. $\begin{bmatrix} 8{,}000 \\ 19{,}100 \\ 13{,}300 \\ 3{,}450 \end{bmatrix}$ represents the revenue by color.

7. C
8. 88; yes
9. $\begin{bmatrix} 3/34 & 4/17 \\ -1/17 & 3/17 \end{bmatrix}$
10. Yes; $AB = BA = I_3$

Quiz Sections 7.3 to 7.5
1. $\begin{bmatrix} 1 & 2 & -3 \\ 2 & -3 & 4 \\ 3 & -1 & 0 \end{bmatrix} \begin{bmatrix} y \\ x \\ z \end{bmatrix} = \begin{bmatrix} 19 \\ -17 \\ 4 \end{bmatrix}$
2. $(x, y, z) = (3, 5, -2)$
3. $(5, -2, 3)$
4. $(1 + 2a, 2 - 3a, a)$
5. $\dfrac{3}{x^2 + 1} + \dfrac{1}{(x^2 + 1)^2}$
6. C
7. $y \leq 4 - x^2$

8.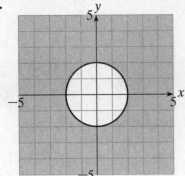

9.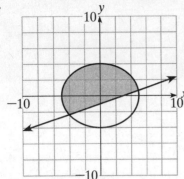

10. C

Chapter Test Form A

1. $(-1, -4), (2, -1)$
2. $(1, 3)$
3. $(-2.20, -7.81), (-0.41, -0.64), (1.11, 5.44)$
4. B
5. $(2, -3, 1)$
6. $(3z + 2, 2z - 1, z)$, where z is any real number
7. D
8. $\begin{bmatrix} 299 & 460 \\ 345 & 207 \end{bmatrix}$ is 1.15 times the original matrix.
9. $\quad 2x + y = 3$
   $-x + 3y + 4z = 0$
   $\quad -2y + z = 5$
10. $(-13, 28, 23)$
11. (a) $\quad x + y + z = 40{,}000$
    $\quad 6x + 8y + 10z = 324{,}000$
    $\quad\quad\quad\; 8y - 5z = 0$

    (b) $\begin{bmatrix} 1 & 1 & 1 \\ 6 & 8 & 10 \\ 0 & 8 & -5 \end{bmatrix} \begin{bmatrix} x \\ y \\ z \end{bmatrix} = \begin{bmatrix} 40{,}000 \\ 324{,}000 \\ 0 \end{bmatrix}$

12. $\begin{bmatrix} 1 & 0 & 0 \\ 22 & 1 & -5 \\ -4 & 0 & 1 \end{bmatrix}$

13. $\dfrac{3}{x + 3} + \dfrac{2}{x - 2}$

14.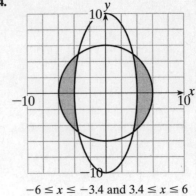

    $-6 \le x \le -3.4$ and $3.4 \le x \le 6$

15. $C = 21$ at $(3, 6)$

Chapter Test Form B

1. $(-1, -5), (2, -2)$
2. $(2, -1)$
3. $(-0.84, -2.68), (0.86, 0.72), (2.31, 3.62)$
4. C
5. $(2z + 4, 3z - 2, z)$, where z is any real number
6. No solution
7. C
8. $\begin{bmatrix} 432 & 360 \\ 240 & 192 \end{bmatrix}$ is 1.2 times the original matrix.
9. $\quad\quad 4y + 3z = -1$
   $\quad 5x + y - 2z = 0$
   $\quad 3x + z = 2$
10. $(1, 2, 2)$
11. (a) $\quad x + y + z = 50{,}000$
    $\quad 5x + 7y + 10z = 376{,}400$
    $\quad 7x - 5z = 0$

    (b) $\begin{bmatrix} 1 & 1 & 1 \\ 5 & 7 & 10 \\ 7 & 0 & -5 \end{bmatrix} \begin{bmatrix} x \\ y \\ z \end{bmatrix} = \begin{bmatrix} 50{,}000 \\ 376{,}400 \\ 0 \end{bmatrix}$

12. $\begin{bmatrix} 1 & 0 & 0 \\ 33 & 1 & -6 \\ -5 & 0 & 1 \end{bmatrix}$

13. $\dfrac{3/2}{x - 3} + \dfrac{1/2}{x + 1} = \dfrac{3}{2(x - 3)} + \dfrac{1}{2(x + 1)}$

Answers **121**

14.

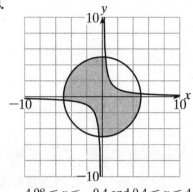

$-4.98 \le x \le -0.4$ and $0.4 \le x \le 4.98$

15. $C = -24$ at $(0, 8)$

Chapter 8
Analytic Geometry in Two and Three Dimensions

Quiz Sections 8.1 to 8.3

1. $(4, -2)$
2. $x = 1/12 y^2 = 0.0833 y$
3. $(x + 4)^2 = 4(y + 4)$
4.

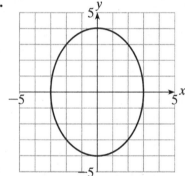

5. A
6. $\dfrac{(x + 1)^2}{16} - \dfrac{(y - 2)^2}{9} = 1.$

7. For an ellipse, the $0 \le e < 1$ and
$e = c/a = \dfrac{\sqrt{a^2 - b^2}}{a}$, so if $e = 0$, $a^2 = b^2$, and

hence the graph of the quadratic is a circle. For all other values of e between 0 and 1, the graph of the quadratic is an ellipse.

8. $(2 \pm \sqrt{39}, -3)$
9. B
10. $\dfrac{(x - 1)^2}{9} - \dfrac{(y + 4)^2}{27} = 1.$

Quiz Sections 8.4 to 8.6

1. Parabola
2.

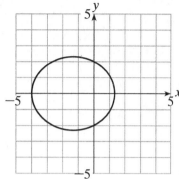

3. Hyperbola; $r = \dfrac{28}{3 + 4 \sin \theta}$
4. $x = -6$
5. $96 > 0$; hyperbola
6. C
7. $(x', y') = (3/2 + 2\sqrt{3}, -3\sqrt{3}/2 + 2)$
$\approx (4.96, -0.60)$
8. C
9. $\sqrt{57}$
10. $x = -1 + 7t, y = 2 - 2t, z = 4 - 7t$

Chapter Test Form A

1. C
2. $x = \dfrac{1}{4}(y + 4)^2 - 4$
3. Center: $(3, 5)$; Foci: $(3, 5 \pm \sqrt{34})$;
Endpoints: $(3, 2), (3, 8)$;
Asymptotes: $y = 5 \pm \dfrac{3}{5}(x - 3)$.
4.

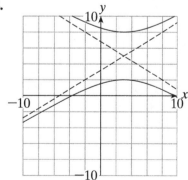

5. $(3, 5)$
6. $x = \dfrac{1}{4}(y - 4)^2 + 2$

7. $r = \dfrac{15}{4 - \sin\theta}$

8. 40 meters

9. Hyperbola: $e = \dfrac{8}{5}$; Directrix: $x = 3$

10. The antenna should be $\dfrac{4}{3}$ feet from the vertex or 16 inches.

11. Hyperbola

12. $y = \dfrac{4 - x \pm \sqrt{25x^2 - 32x - 56}}{6}$

13. $\dfrac{(x-5)^2}{4} + (y+2)^2 = 1$; ellipse

14. $\sqrt{21}$

15. $x = 2 + 3t;\; y = -1 - 5t;\; z = 3 + 3t$

16. $(x-1)^2 + (y-5)^2 + (z+4)^2 = 144$

■ Chapter Test Form B

1. D

2. $y = \dfrac{1}{2}(x+2)^2 - 2$

3. Center: $(-2, 5)$;
Foci: $(-2 \pm 2\sqrt{5}, 5)$;
Endpoints of transverse axis: $(-4, 5)$, and $(0, 5)$;
Asymptotes: $y = \pm 2(x+2) + 5$

4.
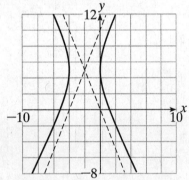

5. $(-1, -10)$

6. $y = -\dfrac{1}{16}(x-2)^2 + 1$

7. $r = \dfrac{15}{4 + \cos\theta}$

8. 52 meters

9. Ellipse: $e = \dfrac{4}{7}$; Directrix: $y = -5$

10. The antenna should be $\dfrac{3}{4}$ feet or 9 inches from the vertex.

11. Ellipse

12. $y = \dfrac{4 + x \pm \sqrt{-23x^2 + 20x + 88}}{6}$

13. $\dfrac{(x-2)^2}{4} + \dfrac{(y+2)^2}{4} = 1$. This is a circle.

14. $\sqrt{14}$

15. $x = 4 + 5t,\; y = -2 - 3t;\; z = 3 + 3t$

16. $(x+3)^2 + (y-5)^2 + (z-2)^2 = 64$

Chapter 9
Discrete Mathematics

■ Quiz Sections 9.1 to 9.3

1. 17,576,000

2. 32,801,517

3. A

4. 0.98

5. $\dfrac{5}{16}$

6. 62%

7. (a) $\dfrac{1}{16}$

   (b) $\dfrac{11}{16}$

8. $3125x^5 - 25{,}000x^4y + 80{,}000x^3y^2 - 128{,}000x^2y^3 + 102{,}400xy^4 - 32{,}768y^5$

9. $508{,}035{,}072\,x^5 y^{15}$

10. D

■ Quiz Sections 9.4 to 9.6

1. Explicit: $a_n = -43 + 15n$
   Recursive: $a_1 = -28$ and $a_n = a_{n-1} + 15$ for $n \geq 2$

2. Explicit: $a_n = \dfrac{1}{8}(2^{n-1})$
   Recursive: $a_1 = \dfrac{1}{8}$ and $a_n = 4 \cdot a_{n-1}$ for $n \geq 2$

3. B

4. -60

5. (a) Converge to $\dfrac{5}{2}$
   (b) Diverge
   (c) Converge to 0

6. 5

7. $P_1 = 1(1!) = 2! - 1$
   $P_k = 1(1!) + 2(2!) + \ldots + k(k!) = (k+1)! - 1$
   P_{k+1}
   $= 1(1!) + 2(2!) + \ldots + k(k!) + (k+1)(k+1)!$
   $= (k+1)! - 1 + (k+1)(k+1)!$

8. $6{,}977.00

9. E

10. $P_1 = 3 = 1(2 \cdot 1 + 1)$

Assume P_k: $3 + 7 + 11 + \ldots + (4k - 1)$
$= k(2k + 1)$

Prove $P_{k+1} = (k + 1)(2(k + 1) + 1)$

$P_{k+1} = 3 + 7 + 11 + \ldots + (4k - 1) + (4(k + 1) - 1)$
$= k(2k + 1) + (4(k + 1) - 1)$
$= 2k^2 + k + 4k + 4 - 1$
$= 2k^2 + 5k + 3$
$= (k + 1)(2k + 3)$
$= (k + 1)(2(k + 1) + 1)$

■ Quiz Sections 9.7 to 9.8

1. Mean = 3.25; Median = 3; Mode = 3, 5

2. Standard deviation = 10.23
Variance = 104.6

3.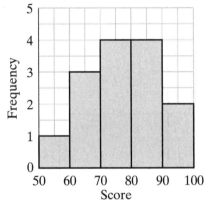

4. {35, 47, 63, 72.5, 77}
Range = 42

5.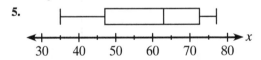

6.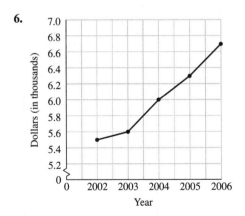

7. D
8. A
9. 86
10. 82

■ Chapter Test Form A

1. (a) 3

(b) $a_n = 3a_{n-1}$

2. C

3. $\sum_{n=1}^{8} 6\left(\frac{1}{3}\right)^{n-1} = \frac{6560}{729} = 8\frac{728}{729} \approx 8.999$

4. (a) $\frac{5}{3}$

(b) Does not converge

5. $32x^5 + 80x^4y + 80x^3y^2 + 40x^2y^3 + 10xy^4 + y^5$

6. $n^3 - n^2 + 5n$

7. 1,256,640

8. 50,400

9. A

10. $\frac{7}{12}$

11. $\binom{6}{r-1} + \binom{6}{r}$

$= \frac{6!}{(r-1)![6-(r-1)]!} + \frac{6!}{r!(6-r)!}$

$\frac{6!}{(r-1)!(7-r)!} \cdot \frac{r}{r} + \frac{6!}{r!(6-r)!} \cdot \frac{7-r}{7-r}$

$= \frac{6![r + (7-r)]}{r!(7-r)!} = \frac{7!}{r!(7-r)!} = \binom{7}{r}$

12.

Stem	Leaf
3	9 7
4	3 8 9 4 1
5	1 7 6 2 6
6	5 1 8 3
7	2 5

13.

Interval	Frequency
30–39	2
40–49	5
50–59	5
60–69	4
70–79	2

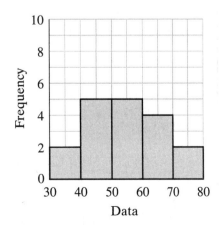

14. Mean: 54.28; Median: 54; Variance: 122.53

15.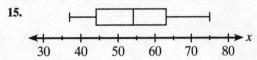

16. P_1: $1(2 \cdot 1 + 1) = 3$

Assume P_k:
$3 + 7 + 11 + \ldots + (4k - 1) = k(2k + 1)$
Prove $P_{k+1} = (k + 1)(2(k + 1) + 1)$

$$\begin{aligned}P_{k+1} &= 3 + 7 + 11 + \ldots + (4k - 1) + (4(k + 1) - 1) \\ &= k(2k + 1) + (4(k + 1) - 1) \\ &= 2k^2 + k + 4k + 4 - 1 \\ &= 2k^2 + 5k + 3 \\ &= (k + 1)(2k + 3) \\ &= (k + 1)(2(k + 1) + 1)\end{aligned}$$

■ Chapter Test Form B

1. (a) -4
   (b) $a_n = a_{n-1} - 4$

2. D

3. $\sum_{n=1}^{27} (-12 + 7n) = 2322$

4. (a) Converges to $\dfrac{5}{2}$
   (b) Does not converge

5. $81x^4 + 108x^3y + 54x^2y^2 + 12xy^3 + y^4$

6. $\dfrac{4}{3}n^3 + 3n^2 - \dfrac{22}{3}n$

7. 3,575,880

8. 1,680

9. E

10. $\dfrac{5}{18}$

11. $\binom{k}{7} + \binom{k}{8} = \dfrac{k!}{7!(k-7)!} + \dfrac{k!}{8!(k-8)!}$

$= \dfrac{k!}{7!(k-7)!} \cdot \dfrac{8}{8} + \dfrac{k!}{8!(k-8)!} \cdot \dfrac{k-7}{k-7}$

$\dfrac{k![8 + (k-7)]}{8!(k-7)!} = \dfrac{(k+1)!}{8![(k+1) - 8]!}$

$= \binom{k+1}{8}$

12.

Stem	Leaf
1	8 9
2	7 9 7
3	3 7 7 8
4	2 5 2 5
5	2 8 6

13.

Interval	Frequency
10–19	2
20–29	3
30–39	4
40–49	4
50–59	3

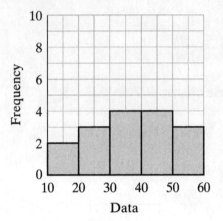

14. Mean: 37.81; Median: 37.5; Variance: 135.03

15.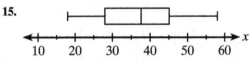

16. P_1: $(3 \cdot 1 - 2) = 1$

Assume P_k: $1 + 7 + 13 \ldots + (6k - 5)$
$= k(3k - 2)$
Prove $P_{k+1} = (k + 1)(3(k + 1) - 2)$

$$\begin{aligned}P_{k+1} &= 1 + 7 + 13 + \ldots + (6k - 5) + (6(k + 1) - 5) \\ &= k(3k - 2) + (6(k + 1) - 5) \\ &= 3k^2 - 2k + 6k + 6 - 5 \\ &= 3k^2 + 4k + 1 \\ &= (k + 1)(3k + 1) \\ &= (k + 1)(3(k + 1) - 2)\end{aligned}$$

Chapter 10
An Introduction to Calculus: Limits, Derivatives and Integrals

■ Quiz Sections 10.1 to 10.4

1. $5(\sqrt{5.3} - \sqrt{4.7}) \approx 0.67$

2. (a) $m = -1$
   (b) $y = -x + 6$

3. C

4. $8x$

5. $[1, 1.5]$, $[1.5, 2]$, $[2, 2.5]$, $[2.5, 3]$, $[3, 3.5]$, $[3.5, 4]$, $[4, 4.5]$, $[4.5, 5]$

6.

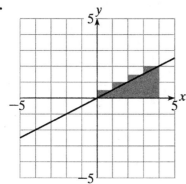

7. 42

■ **Chapter Test Form A**

1. C

2. 30

3. $\dfrac{9086}{5} = 1817.2$

4. 7

5. (a) $-\dfrac{1}{3}$

  (b) $y = -\dfrac{1}{3}x - \dfrac{2}{3}$

6. (a) 64 ft/sec

  (b) 128 ft/sec

7. (a) -32

  (b) $-\dfrac{1}{2}$

  (c) ∞

  (d) -5

  (e) 0

  (f) Does not exist

8. (a) $\dfrac{x^3 + 64}{x + 4}$ is not defined at $x = -4$.

  (b) 48

9. (a) 5

  (b) 4

  (c) Does not exist because $\lim\limits_{x \to 2^-} f(x) \neq \lim\limits_{x \to 2^+} f(x)$.

10. $\{c \mid c < 1 \text{ or } c > 1\}$

11. B

12. 8.75

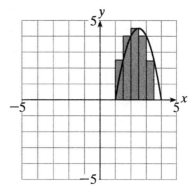

13. $\displaystyle\int_1^2 x^3 \, dx$

14. This area consists of two right triangles, each with legs 2 units long. So the area is $2 \cdot \left[\dfrac{1}{2}(2)(2)\right] = 4$.

15. 12.39

■ **Chapter Test Form B**

1. B

2. -24

3. $-3{,}549.2$

4. -4

5. (a) $\dfrac{3}{16}$

  (b) $y = \dfrac{3}{16}x - \dfrac{9}{8}$

6. (a) 32 ft/s

  (b) 64 ft/s

7. (a) 256

  (b) $-\dfrac{5}{4}$

  (c) $-\infty$

  (d) 5

  (e) 0

  (f) Does not exist

8. (a) $\dfrac{x^3 - 27}{x - 3}$ is not defined at $x = 3$.

  (b) 27

9. (a) 4

  (b) 3

  (c) Does not exist because $\lim\limits_{x \to 4^-} f(x) \neq \lim\limits_{x \to 4^+} f(x)$.

10. $\{c \mid c < 0 \text{ or } c > 0\}$

11. A

12. 4.375

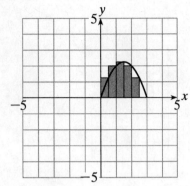

13. $\int_1^3 x^2\, dx$

14. This area consists of two right triangles, each with legs 2 units long. So the area is $2 \cdot \left[\frac{1}{2}(2)(2)\right] = 4$.

15. 11.59

Chapters 6–10
Final Exam

■ Final Exam A

1. $(-1.38, -0.46)$ and $(1.38, 0.46)$
2. $x = 0, y = 4, z = 2$
3. C
4. $(2, 0, -1)$
5. $(5, -8, 7)$
6. $C = 19$ at $(6, 1)$

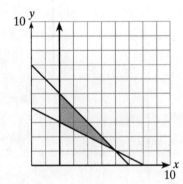

7. $(4, 0)$
8. $\dfrac{(x + 4)^2}{49} + \dfrac{(y + 3)^2}{16} = 1$
9. D
10. $(x - 2)^2 + (y + 1)^2 + (z - 6)^2 = 25$

11. $0 \le \theta \le \pi$

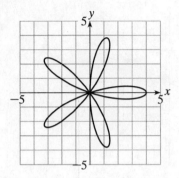

12. B
13. $\langle 253.98, 598.33 \rangle$
14. About 463 ft
15. No, $\langle 2, -1 \rangle \cdot \langle -2, -5 \rangle \ne 0$
16. $3.37°$
17. $x = 2 - 5t, y = -1 - 2t, z = 5 - 5t$
18. $y = \dfrac{2}{x + 3}$; this is a hyperbola.
19. $2\left(\cos\dfrac{17\pi}{15} + i\sin\dfrac{17\pi}{15}\right), 2\left(\cos\dfrac{19\pi}{30} + i\sin\dfrac{19\pi}{30}\right),$
$2\left(\cos\dfrac{49\pi}{30} + i\sin\dfrac{49\pi}{30}\right), 2\left(\cos\dfrac{2\pi}{15} + i\sin\dfrac{2\pi}{15}\right)$
20. $81x^4 - 216x^3y^2 + 216x^2y^4 - 96xy^6 + 16y^8$
21. 6
22. $\$20{,}065.73$
23. $P_1: 1^2 = 1\dfrac{(2(1) - 1)(2(1) + 1)}{3}$

Assume $P_n: 1^2 + 3^2 + 5^2 + \ldots + (2n - 1)^2$
$= \dfrac{n(2n - 1)(2n + 1)}{3}$

Prove
$P_{n+1} = \dfrac{(n + 1)(2(n + 1) - 1)(2(n + 1) + 1)}{3}$
$= \dfrac{(n + 1)(2n + 1)(2n + 3)}{3}$

$P_{n+1} = 1^2 + 3^2 + 5^2 + \ldots +$
$\quad (2n - 1)^2 + (2(n + 1) - 1)^2$
$= \dfrac{n(2n - 1)(2n + 1)}{3} + (2(n + 1) - 1)^2$
$= \dfrac{n(2n - 1)(2n + 1)}{3} + \dfrac{3(2(n + 1) - 1)^2}{3}$
$= \dfrac{n(2n - 1)(2n + 1)}{3} + \dfrac{3(2n + 1)^2}{3}$
$= \dfrac{(2n + 1)}{3}(n(2n - 1) + 3(2n + 1))$
$= \dfrac{(2n + 1)}{3}(2n^2 + 5n + 3)$
$= \dfrac{(n + 1)(2n + 1)(2n + 3)}{3}$

24. $\dfrac{9}{100}$

25. $\dfrac{1}{67{,}600}$

26. $\dfrac{2}{x} + \dfrac{1}{x^2}$

27. D

28. $\dfrac{1}{2}\left(e - \dfrac{1}{e}\right) \approx 1.18$

29. $y = -x + 2$

30. 128 ft

■ Final Exam B

1. $(-1.44, 2.08)$ and $(1.44, 2.08)$

2. $x = 3,\ y = -1,\ z = 1$

3. A

4. $(1, 1, -1)$

5. $(-5, 9, -3)$

6. $C = -21$ at $(9, 3)$

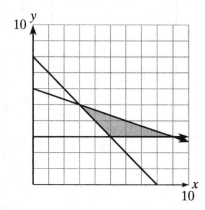

7. $(0, \pm 3)$

8. $\dfrac{(x-4)^2}{16} + \dfrac{(y-3)^2}{49} = 1$

9. B

10. $(x-1)^2 + (y+2)^2 + (z-6)^2 = 9$

11. $0 \le \theta \le 2\pi$

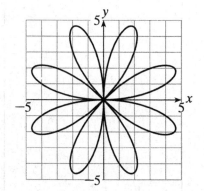

12. C

13. $\langle 579.56, 155.29 \rangle$

14. About 17.86 ft

15. No, $\langle 2, -1 \rangle \cdot \langle -3, -7 \rangle \ne 0$

16. $49.76°$

17. $x = -1 + 4t,\ y = 2 - 3t,\ z = 5 - 5t$

18. $x = \dfrac{3}{y+1}$; this is a hyperbola.

19. $3\left(\cos\dfrac{3\pi}{5} + i\sin\dfrac{3\pi}{5}\right),\ 3\left(\cos\dfrac{19\pi}{15} + i\sin\dfrac{19\pi}{15}\right),$
$3\left(\cos\dfrac{29\pi}{15} + i\sin\dfrac{29\pi}{15}\right)$

20. $x^{10} - 15x^8y + 90x^6y^2 - 270x^4y^3 + 405x^2y^4 - 243y^5$

21. 4

22. $18,424.28

23. $P_1: 1^3 = \dfrac{1^2(1+1)^2}{4}$

Assume $P_n: 1^3 + 2^3 + 3^3 + \ldots + n^3 = \dfrac{n^2(n+1)^2}{4}$

Prove $P_{n+1} =$
$$\dfrac{(n+1)^2((n+1)+1)^2}{4} = \dfrac{(n+1)^2(n+2)^2}{4}$$

$P_{n+1} = 1^3 + 2^3 + 3^3 + \ldots + n^3 + (n+1)^3$

$= \dfrac{n^2(n+1)^2}{4} + (n+1)^3$

$= \dfrac{n^2(n+1)^2}{4} + \dfrac{4(n+1)(n+1)^2}{4}$

$= \dfrac{(n+1)^2}{4}(n^2 + 4n + 4)$

$= \dfrac{(n+1)^2(n+2)^2}{4}$

24. $\dfrac{11}{144}$

25. $\dfrac{1}{17{,}576{,}000}$

26. $\dfrac{2}{x^2} - \dfrac{3}{x}$

27. D

28. $\dfrac{\sin\dfrac{\pi}{2} - \sin 0}{\dfrac{\pi}{2}} = \dfrac{2}{\pi} \approx 0.64$

29. $y = \dfrac{1}{2}x - 2$

30. About 145.45 ft